Locating Value

This book considers the concept of value at the root of our actions and decision-making. Value is an ever-present yet little interrogated aspect of everyday life. This book explores value as it is theorised, practised and critiqued from a variety of disciplinary perspectives. It examines how value is operationalised, endorsed and contested in contemporary society. With international insights from leading scholars, chapters offer a diverse and vibrant geographical engagement with value to showcase its conceptual flexibility. The book explores value's eclectic epistemic foundations; its rollout and legitimation across a range of policy fields; and its challenges and opportunities. The book draws on global examples of value in practice, including forest conservation in Indonesia; protected area management in arctic Norway; a state park in the US; certification schemes for biodiversity in the UK; protection of the international night sky; heritage planning in East Taiwan; a redeveloped airport site in Norway; local food networks in Canada and the UK; a market in the US; and urban development in China.

The book will be of interest to human geographers; political ecologists; heritage scholars and practitioners; planners and those working in public policy; and practitioners and policymakers interested in how valuation processes work.

Samantha Saville is currently an ESRC postdoctoral research fellow in human geography at Aberystwyth University and is extending her doctoral work into value and environmental politics in Svalbard. Her research and publications span interests in polar geography, political ecology, climate change and rural globalisation.

Gareth Hoskins is senior lecturer in geography at Aberystwyth University, where he teaches a variety of topics, including urban geography, the politics of memory, heritage and material culture. His publications involve case studies in the US, the UK and South Africa.

Routledge Studies in Human Geography

This series provides a forum for innovative, vibrant, and critical debate within human geography. Titles reflect the wealth of research that is taking place in this diverse and ever-expanding field. Contributions have been drawn from the main sub-disciplines and from innovative areas of work that have no particular sub-disciplinary allegiances.

For more information about this series, please visit: www.routledge.com/Routledge-Studies-in-Human-Geography/book-series/SE0514

Locating Value
Theory, Application and Critique

**Edited by Samantha Saville
and Gareth Hoskins**

Routledge
Taylor & Francis Group

LONDON AND NEW YORK

First published 2020
by Routledge
2 Park Square, Milton Park, Abingdon, Oxon OX14 4RN

and by Routledge
605 Third Avenue, New York, NY 10017

First issued in paperback 2021

Routledge is an imprint of the Taylor & Francis Group, an informa business

Publisher's Note
The publisher has gone to great lengths to ensure the quality of this reprint but points out that some imperfections in the original copies may be apparent.

British Library Cataloguing-in-Publication Data
A catalogue record for this book is available from the British Library

Library of Congress Cataloging-in-Publication Data
A catalog record for this book has been requested

ISBN 13: 978-1-03-208359-9 (pbk)
ISBN 13: 978-1-138-85223-5 (hbk)

Typeset in Times New Roman
by Apex CoVantage, LLC

Contents

PART II
Spacing value   91

PART III
Practising value   159

Figures

Tables

Contributors

Louise Carver is a researcher at Lancaster University's Environment Centre. Her research considers the political ecology of value and valuation in shifting relations between "naturecultures". Her doctoral research (2013–2017) took place at the Leverhulme Centre for the Study of Value at Birkbeck, University of London. It examined how valuation technologies are designed and used in conservation and biodiversity offsetting strategies. She is also cofounder of The Uplands Projects' rewilding initiative.

Samuel Challéat is a researcher in environmental geography at the UMR CNRS LISST (Toulouse, France). He completed the first French PhD thesis on the emerging problem of light pollution, titled *"Saving the night": Light footprint, urban planning and territorial governance* (2010). Since 2013, Samuel has been coordinating the RENOIR (Ressources environnementales nocturnes and territoires) research group, a collective that brings together perspectives from social sciences and life sciences to examine the night. His current work is on the nocturnal environment, which he strives to build as an object of interdisciplinary research.

Han-Hsiu Chen is an assistant professor in the Department of Cultural Vocation Development, National Taipei University of Technology, Taiwan, where she teaches a range of cultural geography and history courses. Han-Hsiu holds an MA in museum studies from the University of Leicester, and a PhD in human geography from Aberystwyth University. Her doctoral research focused on ongoing heritage practices of tobacco agriculture in Taiwan. Han-Hsui's current work examines the meaning and value transformation of tobacco heritage, and she is also involved in community-led local heritage initiatives in Taipei. She has published numerous academic articles and book chapters on international tobacco heritage practices.

David B. Clarke is professor of human geography and co-director of the Centre for Urban Theory at Swansea University. His research includes work on cinematic cities, consumer culture and value. He is the author of *The Consumer Society and the Postmodern City* (2003); editor of *The Cinematic City* (1997); and co-editor of *Jean Baudrillard: The Disappearance of Culture* (2017), *Jean*

Baudrillard: from Hyperreality to Disappearance (2015), *Jean Baudrillard: Fatal Theories* (2008), *The Consumption Reader* (Routledge, 2003) and *Moving Pictures/Stopping Places: Hotels and Motels on Film* (2009).

Guy Crawford is a PhD student at the Lancaster Environment Centre at Lancaster University, where he works in the department's Political Ecology Group. He holds a BA in sociology from the University of Sussex and an MRes in human geography from Lancaster University. His doctoral work explores the emergence and implementation of a national biodiversity offsetting system in Colombia. He is currently spending time thinking about political ecology, the green economy in practice and state theory.

Tim Cresswell has recently taken up the Ogilvie chair in geography at the University of Edinburgh, moving from Trinity College, Hartford, Connecticut, where he was professor of American studies, dean of faculty and vice president for academic affairs. He holds a PhD in geography from the University of Wisconsin-Madison and a second PhD in English and creative writing from Royal Holloway, University of London. The author of 12 books, Cresswell is an internationally recognized geographer who is intrigued by issues of place and mobility. His most recent book is *Maxwell Street: Writing and Thinking Place* (2019). As a poet, he explores similar themes in his two collections: *Soil* and *Fence* (Penned in the Margins 2013, 2015).

Margarita Díaz-Andreu is a research professor in archaeology at the University of Barcelona. Her work addresses the history of archaeology, identity, heritage and prehistoric art. Her recent interest is in archeoacoustics, and she leads the archeoacoustica research group. She was a member of the Heritage Values Network and is a member of GAPP (Group of Public Archaeology and Heritage). She is the author of several books, has served as an editor for many edited volumes and has published more than 150 articles.

Martin Dijst is a professor of urban geography. He is currently head of the Urban Development and Mobility Department at LISER, Luxemburg. His research is focused on analysing the impact of spatial configurations of land use and transport infrastructures on activity and travel behaviour and on accessibility. He is also interested in mobility, housing and health issues in Chinese cities. Since 2014, he has contributed to studies on urban metabolism, publishing a white paper that develops a combined natural and social science perspective on urban metabolism. His work has been widely published internationally.

Rowan Dixon is a principle professional sustainability and resilience consultant at WSP Opus. His doctoral research with the School of Geography, Environment and Earth Sciences at Victoria University of Wellington investigates reduced emissions from deforestation and forest degradation schemes and commodity chains in Indonesia. He has previously helped Fairtrade and the New Zealand Ministry for the Environment with business and policy advice and development. His research has been featured on the Forests Climate Change Research Hub, Asia Pacific Viewpoint and Environment and Planning A, among others.

Marcus A. Doel is professor of human geography at Swansea University in the UK. He is interested in poststructuralist, post-Marxist and deconstructive geographies, violent geographies, visual culture and the spatial thought of philosophers such as Badiou, Baudrillard, Deleuze and Derrida. His books include *Poststructuralist Geographies: The Diabolical Art of Spatial Science* (1999), *Violent Geographies: Killing Space, Killing Time* (2017), *Jean Baudrillard: Fatal Theories* (2008) and *Moving Pictures/Stopping Places: Hotels & Motels on Film* (2009). He has published over a hundred other works and has written extensively on poststructuralism, human geography and social and spatial theory.

Kalliopi Fouseki is an associate professor at the Institute for Sustainable Heritage of the University College London (UCL). Before joining UCL, she worked at the Acropolis Museum, the archaeological museum of Ancient Olympia and the Science Museum of London. She has also worked as an associate lecturer at the Open University of UK, Greece and Cyprus. Her research interests lie in the areas of heritage and sustainable development, emphasising heritage-led regeneration, heritage in conflict zones, heritage and community participation and heritage values.

Xu Huang is a research fellow at the department of human geography and planning, Sun Yat-Sen University, China. His current research interests are on the topics of belonging to a place, place attachment and neighbourhood social integration. His doctoral research (University of Utrecht, 2015) examined the intersection of value systems, ideology, politics, history and urban planning, with a focus on China. He has coauthored many journal articles on both place attachment and urban planning.

Sjoerd van der Linde is a creative consultant at Studio Louter in the Netherlands, where he delivers communicative museum projects. Previously, he worked as a world heritage specialist at University College London and as assistant professor of archaeological heritage management at Leiden University. His research interests include heritage management, ethnographic research and new media applications in community archaeology. He has also published widely on values and value assessment in archaeology, including a co-edited collection *Fernweh: Crossing borders and connecting people in archaeological heritage management* (2015).

Susan Machum is professor of sociology and the dean of social sciences at St. Thomas University in Fredericton, New Brunswick, Canada. She earned her PhD from the University of Edinburgh, where she studied women's participation in dairy and potato farming in New Brunswick. Having grown up on a small subsistence farm, she experienced firsthand the different value systems embedded in short and long food supply chains. Her experience drives her interest in participatory action research and food value systems. She works to promote the revitalisation of rural communities through her research on sustainability, agriculture, global policy, rural-urban collaboration and the value of local food networks.

Wendy Miller is a postdoctoral researcher for the School of Geography, Earth and Environmental Sciences at Plymouth University. Her research interests and publications include such environmental concerns as energy literacy, sustainability in higher education and research on alternative food networks.

Ana Pereira-Roders is an associate professor in heritage and sustainability at the Eindhoven University of Technology (TU/e) and visiting professor at KU Leuven and ULiege, Belgium. Her research interest is the dual relation between heritage and sustainability in historic urban landscapes. She has particular interest in developing integrated assessment and evaluation frameworks to better monitor and strengthen the conservation and use of heritage worldwide. Ana is widely published in cultural heritage, sustainability and urban studies.

Thomas Poméon is a socio-economist and research engineer at the French National Institute for Agricultural Research (INRA) Observatoire du Développement Rural Unit. He holds a PhD in rural and agro-industrial economy and is involved in research projects on rural development policies and markets related to the interaction between agricultural systems on the one hand and the environment, sustainability and food quality schemes on the other.

Marte Qvenild is a senior adviser at the Research Council of Norway, where she has been working on public perceptions of electricity grids and data sharing. Previously, she was a researcher at the Norwegian Institute for Nature Research (NINA). Her work brings together a range of interests in the socio-environmental field through research into technology, biodiversity, national parks, ecology, invasive species and planning systems. She has published many articles in English and Norwegian high-impact journals.

Gunhild Setten is professor of human geography at the Norwegian University of Science and Technology, where she is deputy head of the Department for Research. She teaches landscape research, geography and the philosophy and history of geography. Her research addresses landscape practices and policies, with a particular focus on the moralities produced by different landscape practices and how different moral notions produce a variety of landscapes. She has worked with farmers and agricultural landscapes, and she has contributed to debates related to cultural heritage management, outdoor recreation and ecosystem services. Her work is widely published in leading international journals and collections.

Joel Taylor is senior project specialist at the Getty Conservation Institute, California. He holds a PhD in historic preservation and conservation from Cardiff

University (2009). Previously, he was a researcher at the Norwegian Institute for Cultural Heritage Research (NIKU), worked at English Heritage and was a lecturer and sustainable heritage MSc course leader at University College London. His work focuses on sustainable heritage, conservation and intergenerationality.

Jan van Weesep is an emeritus professor of social geography from Utrecht University, the Netherlands, and has been knighted for his contribution to Dutch geography. His work has covered issues of social justice, gentrification, housing and the city, which has been widely published.

Preface

In 2014, 'value' seemed to be cropping up everywhere: new journals, institutes, research calls, policy advice. Given this proliferation, we had an inkling that colleagues elsewhere might be working on this. We were grappling with value as an analytical framework and theory and realising the disparate and sparse treatments of the concept in our home discipline of human geography. We thought of how useful it could be to bring together those interested in this area and thus put out a tentative call for papers for the Royal Geographical Society annual conference. We were broadly looking to locate value more firmly in geographical debate and asked potential contributors a series of what we hoped to be provocative questions: How is value produced and reproduced? What roles do (human or nonhuman) stakeholders and the wider material environments play in value's generation and reproduction? What effects do value and/or processes of valuation have (for example on place, community, 'nature', landscape)? How is value experienced, personally and politically, and what can this tell us? How can the concept of value be usefully applied and theorised? We were not sure, if anyone else was really 'out there' beyond our Aberystwyth enclave, thinking along these lines. The abstracts, however, flew in from a surprisingly wide range of directions, and we petitioned successfully for a triple bill on the programme. This collection is a product of that initial call and those interesting conference sessions. We are grateful to all the conference contributors, whether or not they stuck with us through this long-haul project.

In some ways, our concerns for taking value seriously as a powerful concept at work in the world have since been more widely taken up. Value is something we are becoming more used to interrogating, or at least, we're used to being exposed to its operational rollout as private companies and public sector institutions publish (and publicise) their outcomes as metrics for indexes and ranking systems. On the other hand, what has not since emerged is a sustained and broad geographical interest in taking value beyond its traditional and limited conceptual employment. This collection represents the beginnings, we hope, of such a more flexible, diverse and vibrant geographical and interdisciplinary engagement with value. While the individual chapters showcase a range of variously novel and extant approaches to using and examining value, the collection as a whole is greater than the sum of its parts in that it highlights how value and valuation have become ingrained in our everyday lives, producing our realities. We are immensely grateful to the publishers and the stellar cast of international authors for their long-held patience with us in bringing what we think is an insightful collection to a wider audience.

Acknowledgements

Our thanks go to the Royal Geographical Society with the Institute of British Geographers for facilitating our initial call and discussions and to the team at Routledge, who saw the potential in bringing this book to fruition and then stuck with the unexpectedly long project until the end. Likewise, we extend our heartfelt gratitude to all the contributors for continuing to support our aims with this book and who have been extremely patient with us. Samantha Saville also wishes to thank the Economic and Social Research Council for their support for her work, as well as friends and family who have endured many a value-laced rant, especially Stephen Saville.

Of course, behind each chapter are multiple value relations, specifically our contributors, who also have people to thank. The authors of Chapter 3, Kalliopi Fouseki, Joel Taylor, Margarita Díaz-Andreu, Sjoerd van der Linde and Ana Pereira-Roders, would like to thank the JPI-JPHE Pilot Programming on Cultural Heritage and Global Challenge for funding the European project on heritage value (H@V) on which the chapter is based. Wendy Miller (Chapter 4) is grateful for her PhD studentship from Plymouth University. Thanks go to supervisors Geoff Wilson and Richard Yarwood, School of Geography, Earth and Environmental Sciences, for their support, input and guidance. Guy Crawford (Chapter 5) acknowledges the studentship provided by the UK Economic and Social Research Council [ES/J500094/1]. He would also like to thank John Childs, James Fraser, Nils Markusson and Niklas Hartmann for comments on an earlier draft presented to Lancaster Environment Centre's *Work-in-Progress* seminar series for the Society and Environment Research Group. The authors of Chapter 10, Samuel Challéat and Thomas Poméon, extend their thanks to Dr. Tyler Nordgren of the Space Art Travel Bureau for his providing and his permission to use Figure 10.1. Finally, Susan Machum (Chapter 14) acknowledges the research funding from the Canada Research Chairs programme.

1 Locating value

An introduction

Samantha Saville and Gareth Hoskins

This edited volume provides a critical and reflexive consideration of value as it is theorized, applied, and critiqued across a range of disciplinary perspectives, including economic, environmental and cultural geography, heritage and museum studies, political ecology, sociology, and urban and regional planning. The collection explores value's conceptual utility in a wide range of geographical and institutional contexts.

Value's rapidly growing currency in social, cultural, and environmental policy is the latest manifestation of our long-running faith in a concept that has been so central to the philosophy of ethics, aesthetics, and economics as developed through Plato, Kant, Nietzsche, and Marx. In a 21st-century world responding to uncertainty and loss with ever-more-elaborate and varied forms of measurement, value is often regarded as something of a panacea. It can be 'wasted' in ineffective social programmes, not properly accounted for in poor public auditing, and 'accrued' by various techniques of management. This volume examines how value comes to feature in contemporary society and how it is figured, operationalized, endorsed, and contested. It reflects the diversity of value-related critical enquiry that is currently underway in the humanities and social sciences.

Locating Value takes as its starting point the idea that value, whatever else it may be and whatever purpose it serves, is a spatial practice. It is both an abstract form of spacing (by separation, matching, ranking, listing, and grading in comparison on a chart to generate distance and therefore allow exchange through equivalence) and an exercise in physical spacing (a judgement that prompts physical transformation in objects, buildings, places, and landscapes because of the institutional, symbolic, material constellations that judgement generates). Moreover, the act of valuation is "spatially and temporally localized" (Hutter and Stark, 2015, p. 4), and our collective mission is to more effectively locate value: to unpack, explain, and illustrate how value is understood and operationalized. In locating value, we hope to demonstrate the utility of thinking about value geographically. What happens when we think about value as an explicitly spatial phenomenon? What difference does it make to our understanding and theorizing of value if it is recognized as always being located and/or involved in the act of locating? How are certain spaces shaped in the pursuit, or defence, of value and in value's declaration and articulation, and how do those spaces recursively act back on the systems of measurement to help frame what we understand as good, bad, or indifferent?

Rather than work towards a single unifying logic of value, chapters reflect value's conceptual elasticity. They outline value's eclectic epistemic foundations and imperatives; examine value's rollout and legitimation across a range of policy fields; and sketch the contours of its challenge. The scope and flexibility of value as both a scholarly analytic and a governing technology is indicated through the book's coverage of precisely where and how value touches down: forest conservation in Indonesia; protected area management in arctic Svalbard; heritage planning in East Taiwan; local food networks in New Brunswick, Canada, and Plymouth, UK; a redeveloped Norwegian airport site; an open-air street market in Chicago; urban development in Jiangsu Province, China; a state park in the mountains of California; certification schemes for biodiversity offsetting in the UK; and protection of the international night sky.

This introductory chapter outlines our own thinking around this concept and the foregrounding influences and works that have contributed to it. These are not indicative of the wide range of perspectives that the collection represents; there is, we believe, 'value' in the plurality of views we are able to showcase in this book. Moreover, value has been and remains contentious and contested in how we as coauthors have come to know and employ it. This is not smooth, well-trodden intellectual ground, and we have travelled through plenty of potholes and rough terrain in our ongoing discussions.

Why value?

Value is now all-pervasive in formal public conversation. Its popularity in policy-making is due in no small part to its discursive flexibility, providing the appearance of scientific objectivity and technical precision while gesturing to outcomes that extend beyond the economic. In this current discursive moment, value is employed, worried over, and increasingly researched. Government, think tanks, consultancies, charity organizations, and research councils refine methods for calculating value's non-economic expression in terms of nature, health, well-being, heritage, public spirit, and art. The focus of our concern is less about anguished metaphysical questions over what value is ontologically and more about value's epistemics: how value is defined, standardized, practised, mobilized, resisted and made to function in relation to other values. It is about what value comes to mean and how it is legitimized through "data", itself legitimized as "evidence", by those in authority to justify or remove funding. As Helgesson and Muniesa point out, "the performance of valuations are thus not only ubiquitous; their outcomes participate in the ordering of society" (2013, p. 3). Moreover, value is wholly political: "different tropes and deployments of value strike different political points" (Henderson, 2013, p. 34).

Outside policy arenas, value is becoming more prevalent as a form of social critique. Popular books such as Raj Patel's *The Value of Nothing* (2011), Michael Sandel's *What Money Can't Buy* (2012), and Jerry Muller's *The Tyranny of Metrics* (2018) have encouraged us to question how capitalist societies are practising value. We are becoming increasingly used to seeing all manner of companies, charities, institutions, and individuals discussing, declaring, and marketing their values. Since the 2016 UK Referendum on leaving or remaining in the EU, the conversation around what 'British values' are has heightened. The search for value is also consuming considerable resources and intellectual energies.

Value has been key in broad developments of economic and political philosophy. Early philosophical discussions and publications include *The Journal of Value Inquiry*, which has been in print since 1967, and Hartman's *The Structure of Value: Foundations of Scientific Axiology* was published in the same year. Values have been a topic of consideration in cultural studies (e.g. *Cultura: International Journal of Philosophy of Culture and Axiology*) and in discussions on behaviour change and attitudes, particularly in the environmental realm (e.g. the *Environmental Values* periodical). What is interesting is the recent shift from the philosophical to the applied and the related shift from values to *valuation*. One indication of this trend is the founding of the journal *Valuation Studies* in 2013, which specifically covers discussions of valuation as a social *practice* (Helgesson and Muniesa, 2013). To take some UK examples, in 2012 the Leverhulme Centre for the Study of Value was founded and the Arts and Humanities Research Council (AHRC) Cultural Values Project began the following year. This led to further research and eventually a Centre for Cultural Value being established. The Valuing Nature project, funded by the Natural Environment Research Council (NERC) initially, ran from 2011 to 2014 and is likewise continuing to be funded by several UK research councils. Both projects aim to identify and understand value in their respective cultural and natural realms and to "capture" these values via evaluative practices, a recognizably challenging task. After all, can valuations be adequately "robust" as Valuing Nature calls for? Can we really effectively and accurately "capture" it? And if so, who and what has the power and authority to define its limits and evaluative criteria? The chapters in *Locating Value* confront and examine these efforts to capture value. Collectively, we open up the black boxes of often-obscured or unquestioned definitions, assumptions, criteria, and practices that are caught up in such proliferating value practices. First, however, we briefly look at terminology and past works.

Value semantics

Despite the many uses and meanings that "value" is associated with, at the core of value lies the notion of how important something is. Scholars of value concern themselves with how we define what that importance is (Brosch and Sander, 2016). In this vein, Chang proclaims "value is the aim and centre of all human activities and [the] whole [of] human life" (2001, p. 68). While Chang's claim can certainly be refuted, it raises important questions: What is this value, so central yet slippery in meaning? Is it an exclusively human concept? Value is also an everyday term that can take on more nuanced but often-unexplained and often-uninterrogated meanings.

The muddled ways we use the word "value" provide a telling example (Miller, 2008; Skeggs, 2014). The singular article, "value", is often connected to, or synonymous with, a monetary equivalence, or at least taken to be something that can be measured and quantified. We might think of "value for money" or "best value" and the various evaluations and quantifications involved in attributing such labels to an item or service. Conversely, the plural of value does not usually mean multiple quantities. Rather, values are more associated with "subjective feelings", personally held principles by which to live by (Leyshon, 2014) or that which is "held dear" (Lee, 2006; Latour, 2013). Values appear to be qualitative, non-exchangeable and cannot be de-linked from their source, the valuer, or as Miller (2008), describes –in Marxist terminology – values are "inalienable". However, the phrase "value judgement" usually implies a decision based on personal views or standards and not on quantifiable evidence:

> Most people seem blissfully unconcerned with the fact that they use a single term value which can mean both one thing and its very opposite. But what if that is the point? That what value does, is precisely to create a bridge between value as price and values as inalienable, because this bridge lies at the core of what could be called the everyday cosmologies by which people, and indeed companies and governments live?
>
> (Miller, 2008, p. 1123)

It is those everyday cosmologies that we are interested in. Similarly, Roger Lee brings the terms "values" and "value" closer together by examining "social relations of value", that how "people engage in consumption and production and condition the ways in which they come to understand their relationship to the natural and social world" (Lee, 2006, p. 419). Likewise, Bev Skeggs argues that it is the relationships between, and production of, value and values that we should focus on rather than defining exactly what it is/they are. Value, then, to those that would study and analyse its workings at least, becomes a practice, a verb (see Cresswell, this volume); valuations are devised with various component parts and actors assembling to construct calculative devices; valuing, or valuation, is something that is done and performs work (Carolan, 2013) – a social practice (Helgesson and Muniesa, 2013).

Value, as it is acted on and practised, can reveal much about how a system is operating (Raz, 2001). Identifying or detecting and tracing value(s) is a necessary step towards analysing the movements, processes, and contingencies of value. The tools, institutions, and methods of assigning and legitimizing value are fundamental. In analysing value, we need to give due consideration to how value is categorized/measured/judged/"qualculated" (because calculations may not always be quantitative or about pricing, this term incorporates both [Callon and Law, 2003]) and what roles material and human "calculative agencies" (Callon, 1998) play in this work.

The trouble with value

Before we proceed, we will reflect on the difficulties, debates, and challenges that accompany an engagement with value as a theoretical lens or object/subject of critique, some of which are aptly expressed by Ginsberg:

> Value inquiry has loose ends. It is untidy, restless, imperfect, doubt-filled. . . .
> value inquiry offers a model for philosophy as conduct, drawing upon our
> intellectual impulse to continue, and cautioning against our intellectual
> impulse to conclude.
>
> (Ginsberg, 2001, p. 4)

A model philosophy that continues to question does not necessarily sound too
troublesome, but the difficulties of working with value crops up in the writings of
several prominent "value workers". The literary scholar Barbara Herrstein Smith
describes value and value systems as "scrappy" and "discordant and conflictual"
(1988, p. 148). George Henderson in his exploration of Marx's value theory,
writes, "value is very difficult to think" (2013, p. 4). More generously, Michael
Carolan (2013) prefers to think of value as "wild" and not easily fitting into disciplinary structures. To examine where all this anguish comes from, we take a brief
tour around the value landscape.

Value theory, called the labour theory of value, is at the centre of Marxist
ideas. David Harvey's development of this approach is indicative: according to
Doel, "everything hinges on value" (Doel, 2006, p. 55), yet value in Marx is an
obscure and complex concept (Henderson, 2013; Harvey, 2016). Although Marxism is "indispensable" to political economy in enabling generalizations about
capitalism (Christophers, 2014), the focus has been on value production and commodities, with value defined quite specifically through labour. This concentration on production and exchangeable commodities often limits the perspective
to the economic arena (Springer, 2014). Christophers (2014) argues that bringing the performativity of the market into contact with Marxist theory of value
could recoup Marxism's explanatory power by including those missing elements
of what happens when value is exchanged, consumed, and distributed through
markets. Yet capitalist value is thoroughly entangled with social values (Kay and
Kenney-Lazar, 2017), and as the chapters within this volume attest to, there are
many value frameworks and practices that fall outside capitalist market relations:
engaging a strictly Marxist theory of value in exploring our socio-natural world
overly narrows the analysis. If we loosen the ties of value theory from the use-
exchange economic value paradigm, we can open up to working with other theoretical angles such as vital materialism, assemblage, and actor-network theories
(see Fredriksen et al., 2014 for a conceptual mapping review). We can, however,
retain a critical approach and the fundamental principle that value is a social relation (Harvey 2016; Skeggs 2014) rather than a thing, making it all the more difficult to track down, because, to use the famous expression, "value doesn't stalk
around with a sign on its head" (Marx, 2010 [1887]).

Baudrillard's theory of value takes a linguistic turn to posit that values are thoroughly relational in that they rely on opposing terms – for example, beauty and
ugliness – to define themselves. Such systems and moral judgements that come
with them favour the positive side of the opposing pair: beauty in this case (Clarke,
2010). Therefore, as David B. Clarke argues, there is much that value excludes, and
the concept cannot be relied on as a self-sufficient principle, but value is in fact a

"conceptual virus" (2010, p. 235), whereby everything has value, which comes to mean nothing. On the other hand, in value practices that involve ranking, value is a zero-sum game: high worth assigned in one instance means a lower worth assigned elsewhere (see Hoskins 2015). Or as Kay and Kenny-Lazar put it, "the production of value is necessarily accompanied by the production of waste" (Kay and Kenney-Lazar, 2017, p. 301).

Pierre Bourdieu in particular is credited with exposing the processes of assigning cultural value, in his famous essay on forms of social capital (1984). He argued that value as cultural capital is legitimized through upper-class and middle-class citizens having the "appropriate" education, taste, social power, and standing to assign such value for their own (bourgeois) purposes. Rather than artefacts having an inherent value in themselves, they are assigned value for the purposes of legitimizing culture, and value is hence "arbitrary" and thoroughly socially constructed (Bourdieu, 1986, 1993; Skeggs, 2004). However, as Hennion (2015) concludes from his dissection of value processes and constructions in wine tasting, while it is important that seemingly arbitrary valuations receive critical attention, understanding value as totally constructed risks missing both the "something real" that value practices seek to express and the opportunity to engage with how that expression comes into being:

> Tools and procedures, even if fragile, imperfect means, do express something real – the quality of things, however defined, and they grant the competences to make evaluations by confronting and discussing this uneasy, controversial quality.
>
> (Hennion 2015, p. 53)

Indeed, as environmental philosopher Katie McShane has put forward, although intrinsic value may be suspect and difficult to determine, we often behave as if intrinsic value exists (McShane, 2007, 2011). Discussions of values surrounding environmental issues, for instance, can often lead to discomfort, "because value questions are seen as personal, difficult, and basically subjective matters" (Arler, 2003, p. 157). Decision-making, often based on valuations, after all encompasses more-than-rational aspects, such as emotions (Whitehead et al., 2011).

Bourdieu's work also highlights the reciprocal relationship between value as it is declared and the authority of the valuer. The people, systems, or criteria that judge value enjoy status only so far as they make reasonably agreeable valuations. As James English details in his book *Economy of Prestige* (2008), giving awards and prizes for the "wrong" person/film/book can bring the credibility of the judge, award, or their institutional ties into question. The recent overhaul of the Academy Awards (Oscars) to better reflect nonwhite contributions to filmmaking was an attempt to rescue institutional credibility. Numerous individuals refuse British honours (Stephen Hawking, David Bowie, Amartya Sen, Rudyard Kipling, and Peter Benesen [founder of Amnesty International]) for a variety of reasons, including rejecting the status of the monarchy as legitimate. Conversely, creating and shaping awards such as the Nobel prizes, especially for peace and literature, works not only to recognize but to produce a "global moral culture" that expands and is influential even when the judgements and awards given are controversial, contested (Inglis, 2018), and gender biased (Dorling, 2010). For example, awarding the literature prize to Bob Dylan in 2016 caused controversy in literary circles, inflamed by the long wait the committee endured for an acceptance by Dylan of the award and his nonattendance at the ceremony.

As we produce proxies for value in the shape of grades, key performance indicators, awards, and prizes, such measures reflect back on the systems we seek to value. Sometimes, crudely reductive quantification can be a force of broad social good, as in the case of QALY (Quality-Adjusted Life Year) figures employed by the NHS (National Health Service) to defend its limited budget from profiteering pharmaceutical companies hikes in drug pricing. However, results can also be perverse or counterproductive, such as when institutions and individuals seek to game the system (Muller, 2018); for example, in UK education, teaching approaches have shifted towards achieving good results in league tables, and there are many ways schools adapt to perform for OFSTED (Office for Standards in Education, Children's Services and Skills) inspections (Perryman, 2009; Perryman et al., 2018). The higher education sector is a major player in what has become an industry of evaluation: both being subjected to multiple valuations such as the numerous metrics, league tables, and rankings judging research quality, impact, student satisfaction, and many more indices (with at best, questionable results, see Berg et al., 2016) and giving credibility to such systems by participating in and contributing to these assessments and marketing their results.

We evaluate, choose, and make decisions all the time (Ginsberg, 2001), but in socially practising value, we need legitimate ways of categorizing and judging value in some way (Lamont, 2012). This brings us to the very troubles at the heart of the book: the contestations, power relations and practicalities of devising, operating, and living with how we rank, order, and judge what and who matters most. Inevitably, categorization leads to things/people/groups on the border or that defy categorization; producing equivalence between unique entities; faith (or not) in those defining categories; and producing and using the metrological devices in their valuation practices. Yet we argue that despite the difficulties, we should not ignore value or let it pass unquestioned; we need to "stay with the trouble" of value and respond to the knowledges and stories that it produces (Haraway, 2016).

Value in geography

One aim of the book is to make stronger links between the interdisciplinary research that examines value and geographical thinking. Although geographers clearly do analyse everyday relations and value practices, few have explicitly taken a "value lens" for analysis, especially outside of Marxist economic geography and political ecology. Despite being such a central part of human experience and practice, value has largely escaped scrutiny in the discipline as a whole, attention being concentrated largely in isolated silos.

An early exception to the low profile of value in human geography is Burgess and Gold's (1982) edited volume *Valued Environments*, which could have paved the way for a more sustained disciplinary engagement. They set out ideas of value that still resonate today, specifically that value is a dynamic and contingent concept that we should examine: what places are valued and by whom, who assesses value, and how it is assessed. However, besides revisiting ethical questions about how we as geographers conduct ourselves (see the Curry et al., 1996 review of Anne Buttimer's 1974 piece "Values in Geography"), it took another two decades before value was picked up again as a useful analytical tool. An exception to this can be found within the realms of Marxism, political ecology and particularly the combination of the two in studies of neo-liberalizations of nature(s).

Marxist geographers have interrogated the effects that capitalist relations have on the environment we are a part of. Neil Smith and Noel Castree's work stands out here as making significant contributions to this (see Smith's (2009) production of nature thesis and Castree's ongoing work on "social nature" [Castree and Braun, 2001]). Indeed, Castree has periodically explored the progress and trends of examining nature within neo-liberalism (Castree, 2003, 2008a, 2008b, 2010). Much has been achieved empirically in political ecology, human geography, environmental politics, and cognate fields by using such insights, though at times, as Castree (2011) and Bakker (2010) note, without specific mind to precisely identify the varied value processes at work or to make recommendations for policy change or other action (though Apostolopoulou and Adams's (2015) work on land-grabbing has started to address this, and see also Carver, this volume).

Schemes that attempt to value "nature" relative to its utility to the human species are known as "ecosystem services". The advent of these schemes has led to them being an important subject to examine and critique inside and outside of geography (Spash, 2008; Yusoff, 2011; Wynne-Jones, 2012; e.g. Robertson and Wainwright, 2013). Such studies inevitably collide head-on with value, as do parallel schemes in cultural heritage and elsewhere, as many of the chapters in this volume illustrate. Knowledge production and definitions of value, valuation's spatial consequences, and ethnographic accounts of how value frameworks are practised and negotiated all come to the surface through discussions of ecosystems services. The recent collection from the Leverhulme Centre of Value Studies moves these debates on, specifically zooming in on valuation processes in conservation and development (Bracking et al., 2019). Kay and Kenney-Lazar (2017) argue that geographers doing nature-society research should move towards using (still-Marxian-inspired) value as a central framework for analysis rather than neo-liberalism or capitalism, as previous works have tended towards.

Value is beginning to feature more readily in contemporary geographic work, through the widening proliferation of neo-liberal economic practices. Subsequently, economic geographers have built a substantial body of work analysing the global flows of capital. Janelle Knox-Hayes's (2013) research on carbon markets has much to say more generally about the workings of capitalist value. She illustrates how value in use is grounded in a specific space and time with actors and impacts. It is "real" rather than a financialized version: carbon credits, exchanged to represent "nature", or saved emissions, become too abstracted to be successful in increasing the value of our environment. On the other hand, geographers have also sought to draw attention to, understand, and support alternative, diverse economies (the seminal works here are Gibson-Graham, 2008; Gibson-Graham et al., 2013) and currencies (North, 2014) that often work outside, at the margins, or below the radar of capitalist systems and "capitalocentric" thinking.

In relation to economic geography, the transformative processes involved in waste material that changes in social-spatial contexts can trigger from zero, negative, or "rubbish" value to positive and productive value provide rich pickings for geographic observation and analysis. Recent work by Gregson, Crang, and colleagues on global recycling networks and ship-graveyards (for examples, see Gregson et al., 2010 and Crang et al. 2013), uncovers the extent of global trade networks in waste/second-use production materials, disrupting assumptions of the Global South as a dumping ground for the waste of the Global North. It also challenges assumptions of global value chain and network analysis, which has tended to ignore the value in goods and materials post-consumption.

Processes of waste and decay likewise interest geographers thinking about the memories, histories, and stories that decaying or redefined matter can tell (Moran, 2004; DeSilvey, 2006, 2017; Hoskins, 2010). It is here that cultural geography more broadly has picked up on the potential of a value lens. Tim Cresswell's explorations of Maxwell Street are simultaneously interested in the contingency of regimes of value of waste/valuable objects and the practices and processes of valuation at work. Processes of value and valuation are at the heart of constructing archives, the repurposing of "junk" to make musical instruments (Cresswell, 2012), and the revaluation and devaluation of the place itself as it becomes gentrified or erased (Cresswell, this volume). Here, value can be seen as being at the heart of a geographic understanding of the world.

Clive Barnett's work on normativity is related here and draws on some key-shared literatures (such as Dewey's (1939) pragmatism). He calls for normativity to be taken more seriously as an everyday geographical practice and suggests we "develop the analysis of plural geographies of worth" (2014, p. 157). He affirms the importance of practice, evaluation, everyday decision-making, and justification, and he supports the view that such topics have so far been under-researched in geographic investigations. This chimes with the discussions of nature-society geographers in recognising the need for a normative engagement that not only identifies "the specific pathways by which value is extracted from nature" but also addresses "questions about who benefits, who should benefit, and how those imbalances can and should be adjusted" (Kay and Kenney-Lazar, 2017, p. 302). Indeed, geographers have already started to investigate decision-making processes and their spatial consequences from a value(s) perspective, which find (co) productions of knowledge significant in upholding or challenging value regimes (Endres, 2012; Qvenild et al., 2014; Hoskins, 2015).

Despite this eclectic and growing body of work, there has so far not been any pan-disciplinary account of the varied way value can be conceived and articulated. This book provides an interdisciplinary and regionally varied exploration of how value functions and is assigned. Our premise is that an examination of value of this kind could lead to significant political and analytical implications.

Value in action

We think there is conceptual mileage and, indeed, a social imperative to unpack value(s). The philosophical problem of defining value can be a difficult distraction, however. By concentrating on the processes, circulations, measurements, transformations, and representations of value (rather than overly worrying about its definition), we can gain an insight into the work that value does in everyday life and decision-making. Our interest is therefore in how value and values (in their plural forms, regimes, and articulations) are *practised* and what consequences this might have, materially and ethically.

The pragmatist John Dewey began thinking along these lines a century ago. For Dewey, it was important to leave the differences between objective value and subjective value aside and focus instead on what could be empirically observed, the process of valuation itself, the moment of action. He saw this as an ongoing stream of events: "valuations are constant phenomena of human behaviour, personal and associated, and are capable of rectification and development" (Dewey, 1939, p. 57). Therefore, value would be relational and contingent on the circumstances and intent on which the act of judgement takes place. Rorty explains that criteria, a key tool for many valuation practices, are to a pragmatist nothing more than "temporary resting places constructed for specific utilitarian ends" (Rorty, 1982, no pagination). Unpicking what exactly the ends and means are in such contexts, how they are framed, what they effect, how they are negotiated, and how they relate to other value resting places is for us, where much of the work of locating value lies. Located value becomes easier to expose as arbitrary, contingent, and relational; it becomes harder to maintain its pretence of objectivity; and it becomes harder to obscure the way official measures of value claim neutrality.

Whereas many social theorists have now arrived at an agreement of value as a contingent construction, highly dependent on what is being evaluated, by whom, and for what purpose, there are still numerous disciplines and areas of policy that rely on an objective measure of value being possible and useful tools in decision-making. As sociologist Michael Carolan (2013) observes, the relational, contingent, context and practice-dependant nature of value, value creation, and transformation seem to be a given in academic circles. Conversely, for politicians, policymakers, and practitioners, "value is viewed as stable and objectively given" (Carolan, 2013, p. 177).

Investigating these processes, where value is called upon as justification, is political and holds potential for positive change, as Robertson and Wainwright explain well.

> In contesting measure, we challenge the logic by which something becomes a bearer of value in capitalist society; that is, becomes capable of circulating as a means to an end. If we move downstream of this moment and only track

the circulations and chart the injustices and absurdities that result, we have missed the headwaters of analysis and political change.

(Robertson and Wainwright, 2013, p. 900)

It is not only measured, "qualculative", or capitalist value that we may be investigating; by analysing and uncovering the processes of valuation and moments of decision-making, we can gain new perspectives and grasp problems at their source rather than focus on the effects in a more isolated way. Such critiques do not necessarily mean a call to abandon value but are rather a call for transparent discussions on its definitions and uses, for more inclusive negotiation of the whys and hows of criteria and decision-making processes.

> Judgement is often distributed across time and geographical space. It flows, unfolds, and reflects local specificities. It cannot be drawn together at a single commonsense space and time.
>
> (Callon and Law, 2003, p. 4)

As Callon and Law describe of "judgement", value is dispersed, in flux, and has plural meanings depending on the context and relations involved, rather than being static with a fixed definition. In locating value (in its temporary resting places), we can develop an understanding of the roles it plays and "the huge amount of political work that the term performs" (Kay and Kenney-Lazar, 2017, p. 303). It becomes, then, more empirical and less abstract and less a silent conveyer of power.

Structuring value

Although there is much cross-fertilization across the chapters in that they all provide insights into the theorization, spatial nature, and practice of value, we have organized the book into three parts: Knowing Value, Spacing Value, and Practising Value. In Part I, Knowing Value, the first group of chapters focus on the epistemology of value as a means of enquiry and modality of assessment. Our authors provide a range of perspectives and applications of different ways we can approach and locate value in the varied fields it is put to work in. David B. Clarke and Marcus A. Doel start us off with a theoretical exploration into the spatialities contained in Marx's and Hegel's conception of value and the difficulties in tying value down to a concrete definition. If there had been any doubts as to the difficulty of thinking with value in this way, this chapter helps to dispel such doubts. What becomes apparent is that value is inherently geographical: we need to consider "it" within space and time. Kalliopi Fouseki and coauthors do just this in their research into how value is defined and used in heritage professions. Their chapter draws attention to the need to openly discuss what we mean by value in different fields and contexts and how we wish to practise it. Wendy Miller continues this theme of "making values visible". She demonstrates the potential for engaging a capital-assets framework as a tool to assess value that includes human, social, cultural, political, and natural capital and the classical economic capital normally associated with value. Chapter 5 and Chapter 6 dissect the UK's attempt to initiate biodiversity offsetting by devising a valuation metric for wildlife habitats. Guy Crawford analyses the two different discourses of valuing "nature" and socioecological habitats that his research uncovered during the pilot phases of this scheme. Louise Carver's chapter further looks into this scheme through a Marxian examination of the calculative devices used in the process of transforming ecological into economic and creating "value". Together, they demonstrate that value can offer multiple angles of analysis and opportunities for exposing "objective" value as nothing of the sort.

Part II, Spacing Value, moves from considerations of how value can be conceived to how value operates and produces spaces and places. These chapters explore how spatial registers such as place, space, landscape, the city, and the nation, for instance, explicitly inflect notions of value and reciprocally serve as discursive containers through which value gets assigned. Tim Cresswell's selected history of Maxwell Street first locates value in place, connected with the shifting and negotiated values, materialities and biographies of its objects. Second, he reveals a process of value transformation of the place itself via a financial development scheme. Here we see how value practices are enrolled in urban development. Another example of such a process is provided in Chapter 8, by Xu Huanga, Martin Djista and Jan van Weesep. They chart the influence of China's changing political regimes and attendant value systems on urban-planning approaches. The chapter shows how economic and societal changes affect Yangzhou's urban form, changing from an imperial royal city to an ecological garden city over its long history. Marte Qvenild and Gunhild Setten, in their chapter on "snapshot ecology", locate value as a contested and powerful force in the Anthropocene. In their exposition of the fraught process of landscape design and restoration at the former Fornebu Airport, they uncover the temporal and spatial effects of using baseline ecology to categorize species as alien or native. Moving from Oslo suburbs to dark-sky parks and starry night skies, Samuel Challéat and Thomas Poméon bring the discussion of value in conservation to new heights by asking how dark skies are valued. Their chapter details how night skies are being brought into new value frameworks, both monetary and non-monetary, through rural and urban development strategies, protective areas and tourism initiatives. In Chapter 11, Gareth Hoskins looks to the reverse of this process, where protected areas of state parks in California are threatened by closure. His development of a geographical axiology reveals the unsteady and contingent nature of relational value and its inevitable Other: diminishment. The attempts at park closure and their subsequent revaluation push us to further question the underlying politics of valuation at work in much of our society.

Part III, Practising Value, emphasizes examples of value in action, practised and caught in the act. These chapters are concerned with the ways values and value frameworks are entangled in our worldly encounters and variously contested, challenged, stretched, and made malleable. Han-Hsui Chen's chapter works through the ongoing process of identifying and recognizing the heritage of tobacco farming in Taiwan. She links the historical and continuing emergence of Taiwan as independent to the development of different versions and ways of valuing this heritage by different groups for varied political, social, and economic purposes. We stay with the examination of multiple and sometimes conflicting ways of valuing in the following chapter. Samantha Saville analyses the relationship between value and knowledge production and value's subsequent roles in environmental management policies in Svalbard. She argues here for an opening up to value of such discussions and "stakeholder processes" so that more forms of value can be better taken into account. The complexity and multiplicity of value(s) and how different roles and actors conceptualize and practice different versions of it/them is further examined in Susan Machum's piece. In Chapter 14, she argues that simplified "food value chains" hide a complex array of value choices and evaluations at each step. Her chapter also provides insight into how farmers go about matching their own value frameworks to their production and onwards to sale systems. In our final chapter, Rowan Dixon uncovers how value is practised through another scheme designed to value nature, this time through the REDD+ forest schemes of Indonesia. He argues that global value chain analysis can be supplemented to better reflect value in action by including social relations of value. Here we see how the value stories that we tell to operationalize value in different contexts become a contingent part of the "thing"; for example, forest protection schemes can be sold and storied as investment opportunities or corporate social responsibility successes.

To return to one of Haraway's (2016) last books, it matters what ideas and theories we think with and what stories we tell. Value is present and at work in much of our public and academic life. As the examples in this collection show, when value is a framework for analysis, to be discussed, exposed, and located, we gain glimpses of how life could be otherwise. We hope this serves as an inspiration for more geographic work in this vein.

References

Apostolopoulou, E. and Adams, W.M. (2015) 'Neoliberal Capitalism and Conservation in the Post-crisis Era: The Dialectics of "Green" and "Un-green" Grabbing in Greece and the UK', *Antipode*, 47(1), pp. 15–35.

Arler, F. (2003) 'Ecological Utilization Space: Operationalizing Sustainability', in Light, A. and de-Shalit, A. (eds) *Moral and Political Reasoning in Environmental Practice*. Cambridge MA: MIT Press, pp. 155–185. Available at: http://vbn.aau.dk/en/publica tions/ecological-utilization-space(7ad35320-9c2e-11db-8ed6-000ea68e967b).html (Accessed 24 November 2014).

Bakker, K. (2010) 'The Limits of "Neoliberal Natures": Debating Green Neoliberalism', *Progress in Human Geography*, 34(6), pp. 715–735.

Barnett, C. (2014) 'Geography and Ethics III from Moral Geographies to Geographies of Worth', *Progress in Human Geography*, 38(1), pp. 151–160.

Berg, L.D., Huijbens, E.H. and Larsen, H.G. (2016) 'Producing Anxiety in the Neoliberal University', *The Canadian Geographer*, 60(2), pp. 168–180.

Bourdieu, P. (1984) *Distinction: A Social Critique of the Judgment of Taste*. Cambridge, Massachusetts: Harvard University Press.

Bourdieu, P. (1986) 'The Forms of Capital', *Handbook of Theory and Research for the Sociology of Education*. Westport, CT: Greenwood, pp. 241–258.

Bourdieu, P. (1993) *The Field of Cultural Production*. New York: Columbia University Press.

Bracking, S. et al. (eds) (2019) *Valuing Development, Environment and Conservation: Creating Values that Matter*. Abingdon: Routledge (Routledge Explorations in Development Studies).

Brosch, T. and Sander, D. (2016) 'From Values to Valuation: An Interdisciplinary Approach to the Study of Value', in Brosch, T. and Sander, D. (eds) *The Handbook of Value: Perspectives from Economics, Neuroscience, Philosophy, Psychology and Sociology*. Oxford: Oxford University Press, pp. 397–404.

Burgess, J. and Gold, J. (eds) (1982) *Valued Environments*. London: George Allen & Unwin.

Callon, M. (1998) 'Introduction: The Embeddedness of Economic Markets in Economics', in Callon, M. (ed) *The Laws of the Markets*. Oxford: Blackwell Publishers, pp. 1–57.

Callon, M. and Law, J. (2003) 'On Qualculation, Agency and Otherness'. *Centre for Science Studies*. Lancaster University. Available at: www.comp.lancs.ac.uk/sociology/papers/Callon-Law-Qualculation-Agency-Otherness.pdf (Accessed 20 October 2017).

Carolan, M. (2013) 'Doing and Enacting Economies of Value: Thinking Through the Assemblage', *New Zealand Geographer*, 69(3), pp. 176–179.

Castree, N. (2003) 'Commodifying What Nature?', *Progress in Human Geography*, 27(3), pp. 273–297.

Castree, N. (2008a) 'Neoliberalising Nature: Processes, Effects, and Evaluations', *Environment and Planning A*, 40(1), pp. 153–173.

Castree, N. (2008b) 'Neoliberalising Nature: The Logics of Deregulation and Reregulation', *Environment and Planning A*, 40(1), pp. 131–152.

Castree, N. (2010) 'Neoliberalism and the Biophysical Environment 1: What "Neoliberalism" Is, and What Difference Nature Makes to It', *Geography Compass*, 4(12), pp. 1725–1733.

Castree, N. (2011) 'Neoliberalism and the Biophysical Environment 3: Putting Theory Into Practice', *Geography Compass*, 5(1), pp. 35–49.

Castree, N. and Braun, B. (eds) (2001) *Social Nature: Theory, Practice and Politics*. Oxford: Blackwell.

Chang, J. (2001) 'Axiology and Ethics: Past, Present, and Future', in Häyry, M. and Takala, T. (eds) *The Future of Value Inquiry*. Amsterdam: Rodopi, pp. 67–73.

Christophers, B. (2014) 'From Marx to Market and Back Again: Performing the Economy', *Geoforum*, 57, pp. 12–20.

Clarke, D. (2010) 'Value + Structural Law of Value', in Smith, R.G. (ed) *The Baudrillard Dictionary*. Edinburgh: Edinburgh University Press, pp. 234–236.

Crang, M. et al. (2013) 'Rethinking Governance and Value in Commodity Chains Through Global Recycling Networks', *Transactions of the Institute of British Geographers*, 38(1), pp. 12–24.

Cresswell, T. (2012) 'Value, Gleaning and the Archive at Maxwell Street, Chicago', *Transactions of the Institute of British Geographers*, 37(1), pp. 164–176.

Curry, M. R., Lee, R. and Buttimer, A. (1996) *Buttimer, A. 1974: Values in Geography. Commission on College Geography Resource Paper 24*. Washington, DC: Association of American Geographers', *Progress in Human Geography*, 20(4), pp. 513–519.

DeSilvey, C. (2006) 'Observed Decay: Telling Stories With Mutable Things', *Journal of Material Culture*, 11(3), pp. 318–338.

DeSilvey, C. (2017) *Curated Decay: Heritage Beyond Saving*. Minneapolis, MN: Minnesota University Press. Available at: www.upress.umn.edu/book-division/books/curated-decay (Accessed 14 November 2017).

Dewey, J. (1939) 'Theory of Valuation', *International Encyclopaedia of Unified Science*, 2(4).

Doel, M. (2006) 'Dialectical Materialism: Stranger than Friction', in Castree, N. and Gregory, D. (eds) *David Harvey: A Critical Reader*. Oxford: Blackwell Publishing Ltd (Antipode Book Series), pp. 55–79.

Dorling, D. (2010) 'Putting Men on a Pedestal: Nobel Prizes as Superhuman Myths?', *Significance Magazine*, pp. 142–144.

Endres, D. (2012) 'Sacred Land or National Sacrifice Zone: The Role of Values in the Yucca Mountain Participation Process', *Environmental Communication: A Journal of Nature and Culture*, 6(3), pp. 328–345.

English, J.F. (2008) *Economy of Prestige: Prizes, Awards, and the Circulation of Cultural Value*. London, England, Cambridge, MA: Harvard University Press.

Fredriksen, A. et al. (2014) *A Conceptual Map for the Study of Value: An Initial Mapping of Concepts for the Project 'Human, Non-human and Environmental Value Systems: An Impossible Frontier?'* Working Paper 2. University of Manchester: Leverhulme Centre for the Study of Value. Available at: http://thestudyofvalue.org/publications/#2.

Gibson-Graham, J., Cameron, J. and Healy, S. (2013) *Take Back the Economy*. Minneapolis, MN: University of Minnesota Press. Available at: www.upress.umn.edu/book-division/books/take-back-the-economy (Accessed 13 January 2016).

Gibson-Graham, J.K. (2008) 'Diverse Economies: Performative Practices for 'Other Worlds'', *Progress in Human Geography*, 32(5), pp. 613–632.

Ginsberg, R. (2001) 'Value Inquiry as the Future of Philosophy', in Häyry, M. and Takala, T. (eds) *The Future of Value Inquiry*. Amsterdam: Rodopi, pp. 1–6.

Gregson, N. et al. (2010) 'Following Things of Rubbish Value: End-of-life Ships, "Chock-Chocky" Furniture and the Bangladeshi Middle Class Consumer', *Geoforum*, 41(6), pp. 846–854.

Haraway, D. J. (2016) *Staying With the Trouble*. Durham: Duke University Press Books.

Harvey, D. (2016) 'David Harvey Marx & Capital Lecture 1: Capital as Value in Motion'. *First Lecture in the Series: Marx and Capital: The Concept, the Book, the History*, Graduate Center of the City University of New York, 12 September. Available at: http://davidharvey.org/2016/10/david-harvey-marx-capital-lecture-1-capital-value-motion/ (Accessed 6 October 2016).

Helgesson, C.F. and Muniesa, F. (2013) 'For What It's Worth: An Introduction to Valuation Studies', *Valuation Studies*, 1(1), pp. 1–10.

Henderson, G. (2013) *Value in Marx: The Persistence of Value in a More-Than-Capitalist World*. Minneapolis, MN: University of Minnesota Press.

Hennion, A. (2015) 'Paying Attention: What is Tasting Wine About?', in Berthoin Antal, A., Hutter, M. and Stark, D. (eds) *Moments of Valuation: Exploring Sites of Dissonance*. Oxford: Oxford University Press, pp. 37–56.

Herrstein Smith, B. (1988) *Contingencies of Value: Alternative Perspectives for Critical Theory*. Cambridge, MA: Harvard University Press.

Hoskins, G. (2010) 'A Secret Reservoir of Values: The Narrative Economy of Angel Island Immigration Station', *Cultural Geographies*, 17(2), pp. 259–275.

Hoskins, G. (2015) 'Vagaries of Value at California State Parks: Towards a Geographical Axiology', *Cultural Geographies*, 23(2), pp. 301–319.

Hutter, M. and Stark, D. (2015) 'Pragmatist Perspectives on Valuation: An Introduction', in Berthoin Antal, A., Hutter, M. and Stark, D. (eds) *Moments of Valuation: Exploring Sites of Dissonance*. Oxford: Oxford University Press, pp. 1–12.

Inglis, D. (2018) 'Creating Global Moral Iconicity: The Nobel Prizes and the Constitution of World Moral Culture', *European Journal of Social Theory*, 21(3), pp. 304–321.

Kay, K. and Kenney-Lazar, M. (2017) 'Value in Capitalist Natures: An Emerging Framework', *Dialogues in Human Geography*, 7(3), pp. 295–309.

Knox-Hayes, J. (2013) 'The Spatial and Temporal Dynamics of Value in Financialization: Analysis of the Infrastructure of Carbon Markets', *Geoforum*, 50, pp. 117–128.

Lamont, M. (2012) 'Toward a Comparative Sociology of Valuation and Evaluation', *Annual Review of Sociology*, 38(1), pp. 201–221.

Latour, B. (2013) *An Inquiry Into Modes of Existence: An Anthropology of the Moderns*. English-Augmented [Online]. Available at: http://modesofexistence.org (Accessed 13 September 2013).

Lee, R. (2006) 'The Ordinary Economy: Tangled Up in Values and Geography', *Transactions of the Institute of British Geographers*, 31(4), pp. 413–432.

Leyshon, C. (2014) 'Cultural Ecosystem Services and the Challenge for Cultural Geography', *Geography Compass*, 8(10), pp. 710–725.

Marx, K. (2010) *Karl Marx, Frederick Engels Volume 35, Karl Marx – Capital Volume 1*. London: Lawrence & Wishart Electric Book.

McShane, K. (2007) 'Why Environmental Ethics Shouldn't Give Up on Intrinsic Value', *Environmental Ethics*, 29(1), pp. 43–61.

McShane, K. (2011) 'Neosentimentalism and Environmental Ethics', *Environmental Ethics*, 33(Spring), pp. 5–23.

Miller, D. (2008) 'The Uses of Value', *Geoforum* (Rethinking Economy Agro-Food Activism in California and the Politics of the Possible Culture, Nature and Landscape in the Australian Region), 39(3), pp. 1122–1132.

Moran, J. (2004) 'History, Memory and the Everyday', *Rethinking History*, 8(1), pp. 51–68.

Muller, J.Z. (2018) *The Tyranny of Metrics*. Princeton, NJ: Princeton University Press.

North, P. (2014) 'Ten Square Miles Surrounded by Reality? Materialising Alternative Economies Using Local Currencies', *Antipode*, 46(1), pp. 246–265.

Patel, R. (2011) *The Value of Nothing: How to Reshape Society and Redefine Democracy*. London: Portobello.

Perryman, J. (2009) 'Inspection and the Fabrication of Professional and Performative Processes', *Journal of Education Policy*, 24(5), pp. 611–631.

Perryman, J. et al. (2018) 'Surveillance, Governmentality and Moving the Goalposts: The Influence of Ofsted on the Work of Schools in a Post-Panoptic Era', *British Journal of Educational Studies*, 66(2), pp. 145–163.

Qvenild, M., Setten, G. and Skår, M. (2014) 'Politicising Plants: Dwelling and "Alien Invasiveness" in Domestic Gardens', *Norsk Geografisk Tidsskrift – Norwegian Journal of Geography*, 68(1), pp. 22–33.

Raz, J. (2001) 'The Practice of Value', in *Tanner Lectures on Human Values*. Berkeley: University of California. Available at: http://intelligenceispower.com/Important%20E-mails%20Sent%20attachments/The%20Practice%20of%20Value.pdf (Accessed 29 August 2013).

Robertson, M.M. and Wainwright, J.D. (2013) 'The Value of Nature to the State', *Annals of the Association of American Geographers*, 103(4), pp. 890–905.

Rorty, R. (1982) *Consequences of Pragmatism* (1st ed.). University of Minnesota Press. Available at: www.marxists.org/reference/subject/philosophy/works/us/rorty.htm (Accessed 24 November 2014).

Sandel, M. (2012) *What Money Can't Buy: The Moral Limits of Markets*. New York: Farrar, Straus and Giroux.

Skeggs, B. (2004) 'Exchange, Value and Affect: Bourdieu and "The Self"', *The Sociological Review*, 52, pp. 75–95.

Skeggs, B. (2014) 'Values Beyond Value? Is Anything Beyond the Logic of Capital?', *The British Journal of Sociology*, 65(1), pp. 1–20.

Smith, N. (2009) 'Nature as Accumulation Strategy', *Socialist Register*, 43. Available at: http://socialistregister.com/index.php/srv/article/view/5856 (Accessed 31 October 2014).

Spash, C.L. (2008) 'How Much Is That Ecosystem in the Window? The One With the Biodiverse Trail', *Environmental Values*, 17(2), pp. 259–284.

Springer, S. (2014) 'Why a Radical Geography Must Be Anarchist', *Dialogues in Human Geography*, 4(3), pp. 249–270.

Whitehead, M., Jones, R. and Pykett, J. (2011) 'Governing Irrationality, or a More Than Rational Government? Reflections on the Rescientisation of Decision Making in British Public Policy', *Environment and Planning A*, 43(12), pp. 2819–2837.

Wynne-Jones, S. (2012) 'Negotiating Neoliberalism: Conservationists' Role in the Development of Payments for Ecosystem Services', *Geoforum*, 43(6), pp. 1035–1044.

Yusoff, K. (2011) 'The Valuation of Nature: The Natural Choice White Paper', *Radical Philosophy*, 170(Nov/Dec), pp. 2–7.

Part I
Knowing value

2 Spectral geometries

Value *sub specie spatii* and sensuous supersensibility

David B. Clarke and Marcus A. Doel

> It is value . . . that converts every product into a social hieroglyphic. Later on, we try to decipher the hieroglyphic, to get behind the secret of our own social products.
>
> (Marx, 1954 [1887], p. 85)

> But in some sense the damage is done; value sits like a dead letter.
>
> (Henderson, 2013, p. 25)

> The noun "value" may be a dead residue of the verb "value".
>
> (Andrew, 1995, p. 2)

Introduction

"Value", as Karl Marx (1954 [1887], p. 85) memorably observed, "does not stalk about with a label describing what it is". In whatever context this "social hiero-glyph" (Marx, 1954 [1887], p. 85) and "dead letter" (Henderson, 2013, p. 25) appears – aesthetic, economic, moral and so on – both the noun "value" and the verb "value" seem to cloak themselves in deception (Andrew, 1995). All man-ner of antinomies rush in to crowd out conceptual clarity whenever questions of value arise. Is value objective or subjective? Absolute or relative? Concrete or an abstract? Substantial or formal? Does it in fact make sense to frame mat-ters in oppositional terms? Isn't the answer, as with all such putative distinctions, always already "neither"/"both"? Might not value be a "concrete abstraction", for instance? Then again, are we necessarily responsible for imposing such distinc-tions, or are they somehow posited, conjured up or otherwise ushered in by value itself? Whatever the case may be, value persistently threatens to induce a kind of thought-paralysis, against which we are ill equipped to cope. Perhaps something like colour might serve as an apposite analogy, not simply because there is a spec-trum of colours just as there is a spectrum of values but also because colour seems to present much the same set of difficulties as value: is colour objective or subjec-tive; absolute or relative; concrete or abstract; and so on? But would analogical reasoning clarify matters – or end up replicating (or even confounding) the diffi-culties? Alas, even if an analogy with colour were superficially compelling, things might not be so straightforward. Consider, with Stanek (2008, p. 64), Hegel's distinction between the "abstract universal" – "an isolated feature shared by a collection of objects" – and the "concrete universal", which "refers to an essence of a thing considered as embedded into and constitutive of the world of related and interacting things".

Inwood explains the difference between those two types of universal by contrasting *redness* and *life*. Redness is a feature shared by all things red; this feature does not significantly influence the nature of a red thing and its relationships with other red things; thus, it is an abstract universal. By contrast, life, as a concrete universal, "constitutes, in part, an essence of living things, directing their internal articulation, and living things are essentially related to each other in virtue of their life: different species feed off, and occasionally support, each other, and species reproduce themselves" (Inwood, 1992, p. 31 cited in Stanek, 2008, p. 64).

Might value be more like an artificial life-form than a colour? Value, unlike colour, certainly takes on a life of its own. It is common knowledge that value is dynamic, that values rise and fall, that money talks. So, is there an ecology of value rather than a simple spectrum? Or are we in fact dealing with a spectrology and a hauntology (Derrida, 1994; Sprinker, 1999) rather than an ecology, an ecosophy or an ontology? For Arthur, "Capitalism is marked by the subjection of the material process of production and circulation to the ghostly objectivity of value" (2002, p. 216). In that case, the spectre of value *might* materialise as a kind of apparition arising from the "spectrum of . . . connotations" diffracted by the "white light of denotation" (Baudrillard, 1981 [1972], p. 138). Lyotard summons up such a "chromatic ghost", flickering between the material and the immaterial:

> When you want to . . . make a colour film of some little corners in Paris and the suburbs, in November, you realise the inexhaustible dominance of blue-greys in the north European city. All the buildings, all the objects, the faces, the cars, the trees, are ultimately exchangeable (chromatically) into blue-grey. This blue-grey is their *money*, their currency, they can all be converted into that unit. Hear what Klee has to say about grey as a place of *fading away of colours by their balancing*: mutual compensation fund.
>
> (Lyotard, 1998, p. 100, emphasis added)

Value needs its monochromatic blue-grey chequerboard surfaces "to slip in its combinations. . . . It revolves in meta-grey" (Lyotard, 1998, p. 102). Value needs to be spaced out and splayed out.

Given such complexities, we will largely confine ourselves to the analogy between value and (geometrical) space – apart from the occasional linguistic comparison, "for to stamp an object of utility as a value, is just as much a social product as language" (Marx, 1954 [1887], p. 85). Such a geometric analogy was pursued by Marx (2000 [1861–1863]), dissecting the purely formal, relativistic account of value proposed by the anti-Ricardian Samuel Bailey in his *A Critical Dissertation on the Nature, Measures, and Causes of Value* in 1825. In pursuing this, we will inevitably confront value's spectral form, with the occasional blue-grey tinge here and there.

Marx's examination of value *sub specie spatii* ("under the aspect of space") appears straightforward, but it leads ineluctably towards an engagement with that curious ectoplasm that forms the substance of value – or, perhaps, that the form of value secretes as its substance – which can finally be appreciated only by considering Hegel and Kant on space. If this already makes things unwieldy, and one factors in the interminable debate on the relation of Marx to Hegel, matters rapidly become overwhelming. This would be true without adding the importance of understanding Leibniz in order to comprehend Kant on space and adding the fact that the various geometrical treatments in play would dictate the desirability of including Spinoza, particularly in light of the debate staged by Macherey (2011 [1979]). The issues opened up by Marx's geometrical treatment of value call to mind the European Hamlet invoked in Valéry's (1977 [1919], p. 100) *The Crisis of the Spirit*, "bowed under the weight of all the discoveries and varieties of knowledge, incapable of resuming this endless activity", who picks up the skull of "*Leibniz. . . . And this one was Kant, and Kant begat Hegel, and Hegel begat Marx, and Marx begat*". . . In the confines of this chapter, we will have to craft our argument with precision and concision to ensure that as we scratch the surface of value's spatiality, we leave a lasting impression. We therefore intend, following an exposition of Marx's geometrical rendition, to offer some suggestive, scene-setting remarks on its wider context and implications, which will open up some fruitful lines of inquiry, allowing us to put forward one instructive example by way of conclusion: the geometrical value analogy proffered by Simmel (1978 [1907]).

Value *sub specie spatii*

In *Theories of Surplus Value*, Marx (2000 [1861–1863]) considers an argument conceived in terms of space as analogous to (and capable of shedding light on) an argument concerning the nature of value. Targeting Bailey's anti-Ricardian *Critical Dissertation on Value*, Marx finds Bailey guilty of dwelling on the *form* of value ("exchange value" or "money") at the expense of the *substance* of value that Marx saw as crucial (loosely speaking, the productive human labour involved in transforming nature). Bailey, Marx suggests, rightly congratulates himself for putting an end to the search for an invariable standard of value that needlessly occupied Ricardo – but only at the expense of entirely overlooking another, far more fundamental issue. Simplifying matters, Bailey latches onto the idea that money provides a mechanism for comparing (say) chalk and cheese – to the extent that £5 worth of chalk is equal to £5 worth of cheese. To put matters differently, Bailey would not have been fooled by that old schoolyard conundrum: which is heavier, a ton of lead or a ton of feathers? Ricardo's vain search for an invariable standard of value confused matters by failing to be satisfied with the recognition that the *ratio* of lead to feathers would remain unchanged, independently of the yardstick applied. The choice of an imperial ton or a metric tonne to weigh the lead and the feathers will affect the *absolute* quantities of each but not their *relative* proportions. There is no need to discover (nor any meaningful sense of finding) the "right" measure if *relative* proportions are all that matter. Where Bailey errs, according to Marx, is in believing that this puts an end to matters. Whereas the amount of feathers or lead one gets to the ton (or tonne) is determined by nature, by natural law (the law of gravity) – and do not vary in themselves, even if they self-evidently vary in accordance with the unit of measurement selected – the amount of chalk or cheese that one can get for £5 is determined by history, by an *unnatural* law (the law of value). And it can be expected to vary for reasons relating to the history of which they form a (small but perfectly formed) part. In the case of *weighing* feathers or lead, the metric chosen is a purely *formal* concern: the unit of value serves as an arbitrary measure set against an invariable nature. In the case of *valuing* chalk or cheese, however, Marx insists that value necessarily possesses a *substantive* and a formal aspect, its substance relating (if not quite equating) to the socially necessary labour time that it embodies. Thus, "if we could succeed, at a small expenditure of labour, in converting carbon into diamonds, their value might fall below that of bricks" (Marx, 1954 [1887], p. 48). Value is, strictly speaking, weightless – floating, spectral, diffracted – but it is no less materialised and substantiated for that, as we shall see shortly when we encounter the oxymoronic notion of sensuous non-sensuousness.

To hammer his analysis home, pursuing a line of enquiry opened by Bailey, Marx offers a consideration of value *sub specie spatii*. Bailey states,

> As we cannot speak of *the distance of any object* without implying some other object, *between which and the former this relation exists*, so we cannot speak of the value of a commodity but in reference to *another commodity compared with it*. A thing cannot be valuable in itself without reference to another thing any more than a thing can be *distant in itself* without reference to another thing.
>
> (1931 [1825], p. 5)

In response, Marx argues that

> if a thing is distant from another, the distance is in fact a relation between the one thing and the other; but at the same time the distance is something different from this relation between the two things. It is a dimension of the space, it is some length which may as well express the distance of two other things besides those compared.
>
> (2000 [1861–1863], p. 143)

In other words, Marx appeals, on the one hand, to an absolute conception of space: it functions as a container, accommodating two objects set at a distance from one another. Substituting one or the other (or both) objects would not change the distance between them. On the other hand, however, Marx maintains that this does not give us the whole picture:

If we speak of the distance as a relation between two things, we suppose something "intrinsic", some "property" of the things themselves, which enables them to be distant from each other. What is the distance between the syllable A and a table? The question would be nonsensical. In speaking of the distance of two things, we speak of their difference in space. Thus, we suppose both of them to be contained in the space, to be points of the space. Thus, we equalize them as being both existences of the space, and only after having them equalized *sub specie spatii* we distinguish them as different points of space. To belong to space is their unity.

(2000 [1861–18633], p. 143)

Thus, when comparing two objects, we necessarily abstract from their specificity: "we always compare not the 'specific' objects that make one object 'syllable A' and the other 'a table'", says Ilyenkov (1977 [1974], p. 18), "but only those properties that express a third something, different from their existence as the thing enumerated" – in this instance, their location in absolute space (which functions like a two-dimensional or planar scale, with Cartesian coordinates or grid references as the measurements). Likewise, lead and feathers are compared against a third thing – in this case, a more conventional one-dimensional, linear scale – in ascertaining a ton (or tonne) of each. Or again, different geometrical shapes (a parallelogram and a triangle, say) may be said to possess the same area only by abstracting from their other properties.

This is where Marx discovers Bailey's Achilles heel:

if geometry, like the political economy of Mr. Bailey, contented itself with saying that the equality of the triangle and the parallelogram means that the triangle is expressed in the parallelogram, and the parallelogram in the triangle, it would be of little value.

(Marx, 2000 [1861–1863], p. 144)

What is more, value *sub specie spatii* entails something other than Cartesian space, not only because it has a virtual existence but also because "the value dimension is constituted at the very same time as its measure" – which explains why "analogies with other measures such as rulers and weights are very misleading" (Arthur 2004, p. 96 and p. 99, respectively). Value is articulated according to the dialectical play of virtual dimensions and virtual extensions: "for there to be a unity of commodities in a common identity, and determinacy in their relations, they must exist in the same universe and their measure predicated on a common dimension which actualizes their commensurability as values", says Arthur (2004, p. 93). But – crucially – "The value dimension . . . has a purely virtual existence in so far as its reality is merely the ideality of the unity of the commodities in their abstract identity as exchangeable" (Arthur, 2004, p. 96).

When Bailey (1931 [1825], p. 5) says that "a thing cannot be valuable in itself without reference to another thing", Marx insists on considering what that other thing must be, asking, rhetorically, "is social labour, to which the value of a commodity is related, not another thing?" (2000 [1861–1863], p. 143). Marx does not countenance a potentially infinite deferral (dissemination) of self-referential values, like the infinite re-reflection of mirrors that face one another, but insists, instead, on pinning things down, on knowing what is being reflected. Parrying Bailey's claim that "it is impossible to *designate* or *express the value* of a commodity, except by *a quantity of some other commodity*", comes with a crucial qualification, Marx (2000 [1861–1863], p. 146) adds sardonically: "as impossible as it is to '*designate*' or '*express*' a thought except by a quantity of syllables. Hence, Bailey concludes that a thought is – syllables". Marx's point is clear: the one-dimensional form of exchange value is insufficient to fathom value; a second, more substantial dimension is needed. Similarly, when Saussure (1959 [1916], p. 111) sought "to prove that language [*langue*] is only a system of pure values" – demonstrating that language does not simply attach labels to a pre-existing world but provides the grid by means of which the world may be apprehended, just as "the identity of commodities as values is not written on their foreheads" (Arthur, 2004, p. 98) – he attributed two dimensions to linguistic signs. They are always composed of the following (Saussure, 1959 [1916], p. 115):

1   A *dissimilar* thing that can be *exchanged* for a thing of which the value is to be determined
2   A *similar* thing that can be *compared* with the thing of which the value is to be determined

Pressing home the aptness of the term "value" with respect to the sign, Saussure (1959 [1916], p. 115) invokes an economic analogy, suggesting that in order to know what a £5 note is worth, one must grasp both that it can be exchanged for a fixed quantity of a dissimilar thing (bread, say) and that it can be "compared with a similar value of the same system" (£5 is worth five times as much as £1, for example). Marx's "syllables" may well be defined against one another, but this is merely a condition for allowing an exchange of dissimilar substances to take place: thought is not merely a matter of rearranging syllables on the page. The conventional use of particular sounds or marks (signifiers) should not be confused with their linguistic values, any more than the conventional use of certain metals or printed paper as cash should be confused with their monetary values (e.g. a cheque made out to cash is clearly worth more than the paper it is written on). For Saussure, "it is impossible for sound alone, a material element, to belong to language. It is only a secondary thing, substance put to use" (1959 [1916], p. 118). What is proper to language is not, then, sound per se but articulated sound, articulate sound, expressive sound – that "sound-image" dubbed "signifier". Likewise, Saussure stressed that the substance of money is a "secondary thing". So fiat money (such as paper money or electronic money) can function just as well as the substance of money as any precious metal (such as gold, silver or bronze). Marx (1954 [1887], p. 54) concurs: "the value of commodities is the very opposite of the coarse materiality of their substance, not an atom of its matter enters into its composition". Saussure effectively established that meaning is the product of a system of differences – that a *system* of differences is a *condition* for understanding, not that it produces understanding (whether he recognized this or not). This accords with Marx's insistence that Bailey misses something in merely stressing the relative, differential character of value. What that lacuna might be will lead us to Simmel's geometric value analogy, although a comprehensive analysis would demand the triangulation of Simmel's analysis with Marx and Hegel and demand a fuller appreciation of Kant and Leibniz, to whom we now turn.

Sensuous supersensibility

The foregoing discussion of value *sub specie spatii* implicitly bears the weight of an intensely heated and protracted debate over the fundamental nature of space: a debate that occupied both Kant, partly in response to Leibniz, and Hegel, partly in response to Kant. Given the importance of Hegel to Marx, it is unsurprising that Marx's comments concerning the "coarse" material substance of value, or rather its absence from value, should bear on this debate. When Lenin (1976 [1914–1916], p. 172) pronounced that "Hegel is essentially *right*: *value* is a category which *enthbehrt des Stoffes der Sinnlichkeit* [dispenses with the material of sensuousness]", the debate over the nature of space remained implicit. Rendering it more explicit, Lenin cites Hegel (1969 [1816], p. 1306) – "could it have ever been thought that philosophy would gainsay the validity of the intelligible essences because they are without the spatial and temporal material of sensuousness?" – to side with Hegel over Kant, contending that abstraction "does not get away *from* the truth but comes closer to it" (Lenin, 1976 [1914–1916], p. 171). Marx, using the example of a wooden table, says that "an ordinary sensuous thing [*ein ordinäres, sinnliches Ding*]", once it has become a commodity, is "metamorphosed into a supernatural thing, a *sensuous non-sensuous* thing, sensuous but non-sensuous, sensuously supersensible (*verwandelt er sich in ein sinnlich übersinnliches Ding)*" (cited in Derrida, 1994 [1993], pp. 188–189). As Derrida notes, this "literally recalls (and this literality cannot be taken as fortuitous or external) the definition of time – of time as well as of space – in Hegel's *Encyclopedia (Philosophy of Nature, Mechanics)*" (1994 [1993], p. 194), which Derrida proceeds to quote (cf. Hegel, 1970 [1830], p. 1: §258). Malabou, citing this passage from Hegel, highlights the significance of Hegel's relation to Kant:

> Time is in fact presented *at once* according to its classical Greek determination, that of Aristotle, and according to its modern determination, that of Kant. If the analysis of the now, the definition of time as "a being which in being, is not" (Aristotle 1984 IV, 10, 218 b 29) is effectively borrowed from *Physics* IV, the definition of time as "the pure form of sensibility" – Hegel writes, in fact: "time, like space, is a pure form of sense or intuition; it is the non-sensuous sensuous (*das unsinnliche Sinnliche)*" (Hegel 1970, p. 1 and p. 230; 1969–1979, p. 9 and p. 48) – is clearly taken from the *Critique of Pure Reason* (Kant 1996).

> (Malabou, 2000, p. 209, original emphasis)

Malabou's reference to Aristotle pertains to the Now, which, in being, has to pass and be no more: like a point, it has position but no extension. For Aristotle, the Now is always the same yet also the "non-same", always in a different position: a "same non-same". The sense in which Hegel's definition is, as Malabou says, "taken from" Kant is hardly straightforward.

The crucial point here concerns how space, like time, amounts to the "non-sensuous sensuous" in Kant and Hegel and why Marx should invoke this quality in relation to the commodity and the nature of value. We can note, initially, that although it is the conception of space in Kant's (1781) *Critique of Pure Reason* that is of direct relevance, Kant's precritical treatment of space – specifically the argument from "incongruent counterparts" first presented in Kant's (1992 [1768]) *Regions in Space* essay – has been seen as crucial to the development of Kant's transcendental philosophy. That work, while not supporting Newton's substantialist account of *absolute* space, nevertheless levelled serious objections against Leibniz's *relative* conception of space (Clarke and Leibniz, 1956 [1717]). Incongruent counterparts are forms that in two-dimensional space, cannot be superimposed on their mirror images through rotation (for example, while the letter E has a congruent counterpart, Ǝ, the letter F has an incongruent counterpart, Ⅎ). For Kant, the existence of incongruent counterparts demonstrates that space possesses properties that cannot be reduced to *relative* spatial configurations alone. This is exemplified in three-dimensional space by left and right hands: the relations between any set of points on my left hand would be identical to the corresponding relations on my right hand as a set of internal differences, but this would never make a right-hand glove fit my left hand. This property of "handedness" – chirality – entails that space possesses qualities that are not apprehensible in terms of conceptual difference alone (this is *this* because it is not *that*). For Deleuze (2004 [1968], p. 15), "it is Kant who best indicates the correlation between objects endowed with only an indefinite specification, and purely spatio-temporal or oppositional, non-conceptual determinations (the paradox of symmetrical objects)".

Whether or not it provided the "key to transcendental idealism" (Buroker, 1981, p. 133), the argument from incongruent counterparts is crucial to the development of Kant's critical treatment of space as a pure form of intuition, free of any necessary empirical content. Kant's Copernican Revolution saw space as empirically real but transcendentally ideal: without the "*purely sensory*, the formal sensory free of all sensual matter . . . no Copernican Revolution would have taken place" (Derrida, 1982 [1972], p. 44). Kant deployed a strategic reversal. Just as Copernicus reversed the stultifying assumption that celestial bodies circled the observer to see if greater success might accrue "if he made the observer revolve and left the stars at rest", so did Kant, rather than resting on the assumption that "our thoughts must conform to . . . objects", reverse this to proceed on the basis that "objects must conform to our cognition", in order to discover means of determining "something about objects before they are given to us" (Kant, 1996 [1781], Preface, B: xvi). "Before it had all seemed so simple: things cast shadows", Krzhizhanovsky (2006, p. vii) pronounces: "but now it turned out that shadows cast things".

In the *Critique of Pure Reason* – cited here adopting the convention of using A and B to indicate the first and second editions – Kant (1996 [1781]) ranges over terrain subsequently traversed by Marx (2000 [1861–1863]) in considering value *sub specie spatii*. Asking "what, then, are space and time? Are they real existences? Are they only determinations or relations of things, yet such as would belong to things even if they were not intuited?" (Kant, 1996 [1781], A, p. 23; B, p. 37), Kant determines that "space is not an empirical concept which has been derived from outer experiences" (B, p. 38) but rather "the subjective condition of sensibility, under which alone outer intuition is possible for us" (A, p. 26; B, p. 42). Space, like time, is crucial to "the possibility of a priori synthetic judgments, the very possibility of integrating what I sense with what I conceptualize" (Bass, 2014, p. 35).

Hegel's engagement with space is embroiled in a broader rejection of Kant's transcendental philosophy, which is seen as limiting the scope of knowledge by focusing unduly on the mental capacities of the subject (famously likened by Hegel as refusing to enter the water until one has learned to swim). The crucial point is that

> Hegel believes that when Kant asks in the [Transcendental] Aesthetic *What are space and time?* and answers that they are pure intuitions, his answer is worse than incomplete, even in the context of the *Critique*. The space of which we ask *What is it?* is an object of experience, and as such, a product of synthesis in accordance with the categories.
>
> (Jenkins, 2010, p. 341, original emphasis)

For Hegel, the pure intuition that Kant equates with space is, "at best, one necessary condition of the existence of space, and Kant has not provided an answer to the question of what space might *be* (or, as Hegel would put it, he has not provided its concept)" (Jenkins, 2010, p. 341).

Addressing the nature of space, Hegel attempts to derive it from being: "the primary or immediate determination of nature is the abstract universality of its self-externality, its unmediated indifference, i.e., space" (Hegel, 1970 [1830], p. 1, p.§254). In the *Phenomenology of Spirit*, Hegel refers to the Here – in the context of an examination of regions in space as parts of a whole – as a universal, which he defines as that "which *is* through negation, which is neither This nor That, a *not-This*, and is with equal indifference This as well as That" (Hegel, 1977 [1807], p. 60 and p. 96). By implication, in relation to space, the Here is also a not-Here: "an individual Here is an entity that vanishes or cancels itself [*ein Aufgehobenes*]", as Jenkins (2010, p. 339) puts it. Like the Now in time, "the *Here pointed out*, to which I hold fast, is similarly a *this* Here which, in fact, is *not* this Here, but a Before and Behind, an Above and Below, a Right and Left" (Hegel, 1977 [1807], p. 64 and p. 108). It is worth citing Jenkins at length on this point:

> Hegel often asserts that space is real, or exists, only as *determined* – that is, as ordered and unified by concepts, or by spontaneity in general. In the

Encyclopedia Logic, for example, he asserts that "space and time are actual only through their determinacy (i.e., as *here* and *now*), and this determinacy lies in their concept" [Hegel, 1975 [1817], p. 79]. This remark should bring to mind Hegel's complaint in his lectures that Kant does not approach the topic of space and time as he should, by asking *What is their concept?* but instead by asking *Are they external things or something in the mind?* [Hegel, 1996 [1840], p. 587]. The same point appears in the *Phenomenology* when Hegel notes that "this Here" is negative, insofar as "negativity" is a general term for determination. Hegel's implicit suggestion is that if Kant had asked the former question concerning the concept of space, he would not have arrived at his position on the subjectivity of space. As determined, or essentially articulated, space is not a mere form of intuition within us but something we confront in experience.

(2010, p. 340, original emphasis)

We can add little more to this here, beyond stressing its importance in terms of the striking parallel with value, which we shall discuss briefly in relation to Hegel before we make some more direct but necessarily succinct comments on Marx's theory of value vis-à-vis Hegel.

Hegel deployed the notion of value across a diverse range of contexts – political economic; moral; aesthetic; etc. – drawing on a conception of measure that, intriguingly, derives from the chemistry of his day, specifically, the "very special sense of measure understood as the measure of atomic masses and the measure of 'stoichiometric proportions' (the proportions defining specific compounds)" (Deranty, 2005, pp. 312–313). Thus, Hegel draws on the "Measure" section of the *Encyclopaedia Logic* (1975 [1817]), in his lectures on the philosophy of right, to define value as "an abstraction and the transformation of the qualitative into the quantitative, whereby the qualitative fully retains its right, and determines the *quantum*" (Hegel, 1974 [1822–1823], p. 240, cited in Deranty, 2005, p. 311). Rather than a reductive transformation of qualities into quantities, the parallel provided by this particular sense of measure allows Hegel to argue that

value is indeed a quantity, but a quantity that expresses the qualitative difference of entities in their commensurability to all other entities. . . . Particular measuring by definition requires the background of the commensurability of all measurable things, measuring is also co-measuring.

(Deranty, 2005, p. 315)

Arthur's (1988, 1993, 2004, 2014) sustained examination of the debt to Hegel in Marx's value theory offers the fullest translation of how exchange value serves as a "measure" to generate an account of value as "pure quantity".

The vindication of my category of "value as pure quantity" follows from the fact that the quantity of exchangeables has no inherent limit. Every exchangeable relates to putatively infinite others. Equally, the many, considered as determinate, consists of discrete "ones". Every "one" has to be determined as an exchangeable item if exchange is to be possible. It is not enough for the commodities to be specified as having properties that make them exchangeable in a general indeterminate sense; a determination is required that allows for discrete exchangeables to be presented for exchange. In other words a commodity must be specifiable as an item for sale. It has to specify itself in discrete units, each of which – the quantum – announces itself as an instantiation in *delimited* form of the good concerned. A baker does not sell "bread" but a loaf of such and such a weight. Only thus does sale become determinate.

(Arthur, 2014, p. 273)

How value is expressed in the commodity is, as Arthur demonstrates, responsible for crystallising the sensuous and the supersensuous. This may be appreciated superficially by noting that, in a simple example where one use value is expressed in terms of another – for example, z of commodity A expresses its value in y of commodity B – "value, defined as not-use-value A, is given in use-value B, so it is supersensuous and sensuous at the same time, but in relation to two *different* commodities" (Arthur, 2014, p. 283). As a fuller analysis would show, in relation to the commodity and the functioning of a universal equivalent in the form of money,

value, originally defined in opposition to the use-value of A (hence a supersensuous reality), *is* now use-value A (a sensuous reality). . . . The two worlds of value, the sensuous and supersensuous, are here immediately one; the very same commodity contains both worlds. They are essentially related. The commodity is "a sensuous supersensuous thing".

(Arthur, 2014, p. 283)

For Derrida, "the commodity thus haunts the thing, its spectre is at work in use-value" (1994 [1993], p. 153). Use value is spirited away by exchange value *from the off.* They come together – or not at all. They are two sides of the same coin.

This unavoidably condensed argument raises an immediate issue: the extent to which Marx – who by his own acknowledgement, "in the chapter on the theory of value, coquetted with the modes of expression peculiar to [Hegel]" (Marx, 1954 [1887 Afterword to the German edition], p. 25) – secreted something of the "mystifying side of Hegelian dialectic" into his analysis. As Derrida notes, "Marx does not like ghosts any more than his adversaries do. He does not want to believe in them. But he thinks of nothing else" (1994, pp. 46–47). If, however, Marx remained haunted by the grotesque thoughts spun out of the inanimate brains of commodities, for Bauman (1978, p. 54) such preoccupations simply acknowledge the alienation wrought by the topsy-turvy world of value rather than enact back-to-front thinking: "mystification is the truth of reality as long as reality is estranged". Indeed, "all that palaver about the necessity of proving the concept of value comes from complete ignorance both of the subject dealt with and of scientific method" (Marx, 1977 [1868], p. 524).

Conclusion

Having offered an overview of the issues that the juxtaposition of space and value brings to the fore, I must still demonstrate how a spatial analysis might clarify the nature and operation of value. In an exposition that resonates with the Hegelian conception taken up by Marx, the geometrical analogy between the value of an object and the length of a line considered by Simmel in *The Philosophy of Money* (1978 [1907]; we cite from a superior translation, extracted in an earlier anthology) offers an apt illustration. Simmel's focus is on the relativity of value – treading similar ground to that covered in the first section of this chapter.

If economic transactions generally entail that equal values are exchanged, the temptation is to assume that the value of an object must have been established before entering into exchange: "it would seem that two things can have the same value only if each of them already has its own value", ventures Simmel (1971 [1907], p. 50). This would also seem to be "upheld by the analogous argument that two lines can be equally long only if each of them possesses a determinate length before the comparison". But, Simmel notes, "if we look at the matter closely . . . we see that a line possesses this length only at the moment of being compared with another line". If this seems counterintuitive, Simmel – echoing Bailey's (1931 [1825], p. 5) remarks that a thing cannot be "*distant in itself*" – notes that "A line is not 'long' of and by itself" (Simmel, 1971 [1907], p. 50). While one might note that a line, unlike a point, has extension by definition, Simmel's argument is that

> it cannot determine its length by itself, but only through another line by which it is measured, and which it measures as well, although the *result* of the measuring is not determined by the process of the measuring, but depends on each of the two independent lines.
>
> (1971 [1907], p. 50)

(Even then, as Arthur (2004, p. 96) reminds us, to announce "that this is equivalent in length to that, through laying them side by side, does not in itself give a measure of either".) Making a comparison with value judgements, where the desire to judge things creates something that does not belong to the things yet is prompted by our relation to them, Simmel notes:

> The same is true of judgements of length. The *demand* to make such a judgement emanates, as it were, from things, but the content of the judgement is not indicated by the things; it can only be realized through an act within ourselves. That length is not contained in the individual object but arises out of a process of comparison is easily hidden from us, because from the individual instances of relative length we have abstracted the universal concept of length – from which the *determinacy* that is indispensable for any concrete length is excluded. We then project this concept back into things, and suppose that they must originally have had *length* even before this could be determined in the individual case through comparison.
>
> (Simmel, 1971 [1907], pp. 50–51)

The universal measure appears to be the simple opposite of a relative comparison, but this is not the case:

> Out of numerous individual comparisons of length fixed measures are crystallized which are then used to determine the length of all spatial figures, such that these measures, the embodiment, as it were, of that abstract concept of length, seem removed from relativity, since everything is measured by them but they are not themselves measured. To think this is to commit an error no less egregious than to think that the falling apple is attracted by the earth, but not the earth by the apple.
>
> (Simmel, 1971 [1907], p. 51)

If this argument is compatible with the Hegelian lineage feeding into Marx's value theory, it is, again, a spatial analysis that lends value greater clarity. Yet there remains a serious problem that, if only as a parting shot, requires noting – a problem concerning the relations between space and time and the tendency, originating with Aristotle, to spatialise time (Bergson, 1970 [1889]; 1999 [1922]): to conceive of time in terms of a linear spatio-temporal unity. Indeed, Heidegger (1962 [1927], p. 49), speaking of Aristotle, says that "every subsequent account of time, including Bergson's, has been essentially determined by it". This, for Derrida, discloses a fundamental problem of dialectical thought (including as it underpins Marx's theory of value):

> Aristotle . . . defines time as a dialectic of opposites, and as the solution of the contradictions that arise in terms of space. As in the *Encyclopedia*, time is the line, the solution to the contradiction of the point (nonspatial spatiality). And yet it is not the line, etc. The contradictory terms posited in the aporia are simply taken up and affirmed together in order to define the *physis* of time. In a certain way, one might say that dialectics only always repeats the exoteric aporia by affirming it, by making of time the affirmation of the aporetic.
>
> (1982 [1972], p. 54)

The sense in which, "according to a fundamentally Greek gesture, this Hegelian determination of time permits us to think the present, the very form of time, as eternity" (Derrida, 1982 [1972], p. 45) hinges on a metaphysics that sees space as inhabiting the present: Here and Now. Insofar as the time responsible for hundreds of millions of years of evolutionary history is thereby put on an equal footing with the "leap second" applied to Coordinated Universal Time (UTC) on 30 June 2015 at 23:59:60 UTC, one finds parallels in how "Ricardo and his predecessors confused the universal character of labour (as the source, along with nature, of wealth in all societies) with its particular characteristic under capitalism, as the creator of value" (Pilling, 1980, p. 47). When, following Adam Smith, Ricardo "makes the primitive hunter and the primitive fisher straightway, as owners of commodities, exchange fish and game in the proportion in which labour-time is incorporated in these exchange-values", then, as Marx (1954 [1887], p. 76) sardonically observes, Ricardo "commits the anachronism of making these men apply to the calculation, so far as their implements have to be taken into account, the annuity tables in current use on the London Exchange in the year 1817". Sadly, most of us continue to address value anachronistically. The critique of political economy *sub specie spatii* inaugurated by Marx remains to be accomplished.

References

Andrew, E.G. (1995) *The Genealogy of Values: The Aesthetic Economy of Nietzsche and Proust*. Lanham, MA: Rowman and Littlefield.

Aristotle (1984) 'Physics', in Barnes, J. (ed) *The Complete Works of Aristotle*. Princeton, NJ: Princeton University Press.

Arthur, C.J. (1988) 'Hegel's Theory of Value', in Williams, M. (ed) *Value, Social Form and the State*. London: Macmillan, pp. 21–41.

Arthur, C.J. (1993) 'Hegel's *Logic* and Marx's *Capital*'. in Moseley, F (ed) *Marx's Method in Capital: A Re-examination*. Atlantic Highlands, NJ: Humanities Press International, pp. 63–88.

Arthur, C.J. (2002) 'The Spectral Ontology of Value'. in Brown, A., Fleetwood, S. and Roberts, J.M. (eds) *Critical Realism and Marxism*. London: Routledge, pp. 215–233.

Arthur, C.J. (2004) *The New Dialectic and Marx's* Capital. Leiden: Brill.

Arthur, C.J. (2014) 'Marx, Hegel and the Value-form'. in Moseley, F. and Smith, T. (eds) *Marx's* Capital *and Hegel's* Logic: *A Re-examination*. Leiden: Brill, pp. 269–291.

Bailey, S. (1931 [1825]) *A Critical Dissertation on the Nature, Measures and Causes of Value; Chiefly in Reference to Mr. Ricardo and His Followers*. London: London School of Economics Series of Reprints of Scarce Tracts in Economic and Political Science, p. 7.

Bass A. (2014) 'The Signature of the Transcendental Imagination'. *The Undecidable Unconscious: A Journal of Deconstruction and Psychoanalysis*, 1, pp. 31–51.

Baudrillard, J. (1981 [1972]) *For a Critique of the Political Economy of the Sign* (trans. Levin, C.). St Louis MO: Telos.

Bauman, Z. (1978) *Hermeneutics and Social Science: Approaches to Understanding*. London: Hutchinson.

Bergson, H. (1970 [1889]) 'Aristotle's Concept of Place', in *Studies in Philosophy and the History of Philosophy* (Vol. V). (trans. Ryan, J.K.). Washington, DC: Catholic University of America Press, pp. 13–72.

Bergson, H. (1999 [1922]) *Duration and Simultaneity: Bergson and the Einsteinian Universe* (ed. Durie, R. and trans. Jacobson, L.). Manchester: Clinamen.

Buroker, J.V. (1981) *Space and Incongruence: The Origin of Kant's Idealism*. Dordrecht: Reidel.

Clarke, S. and Leibniz G. W. (1956 [1717]) 'The Leibniz – Clarke Correspondence: Together With Extracts from Newton's'. in Alexander, H.G. (ed) *Principia and Optiks*. Manchester: Manchester University Press.

Deleuze, G. (2004 [1968]) *Difference and Repetition* (trans. Patton, P.). London: Continuum.

Deranty, J.P. (2005) 'Hegel's Social Theory of Value'. *The Philosophical Forum*, 36(3), pp. 307–331.

Derrida, J. (1982 [1972]) 'Ousia and grammē: Note on a Note from Being and Time', in *Margins of Philosophy* (trans. Bass, A.). Chicago, IL: University of Chicago Press, pp. 29–67.

Derrida, J. (1994 [1993]) *Specters of Marx: The State of the Debt, the Work of Mourning, and the New International* (trans. Kamuf, P.). London: Routledge.

Hegel, G.W.F. (1969 [1816]) *Science of Logic* (trans. Miller, A.V.). London: Allen and Unwin.

Hegel, G.W.F. (1969–1979) *Werke in zwanzig Bänden*. Frankfurt am Main: Suhrkamp.

Hegel, G.W.F. (1970 [1830]) *Philosophy of Nature: Part 2 of the Encyclopedia of Philosophical Sciences* (trans. Petry, M.J.). London: Allen and Unwin.

Hegel, G.W.F. (1974 [1822–1823]) *Vorlesungen über Rechtsphilosophie* [*Lectures on the Philosophy of Right*] (Vol. 3). (ed. Ilting, K-H.). Stuttgart: Frommann Verlag.

Hegel, G.W.F. (1975 [1817]) *Logic: Part I of the Encyclopedia of the Philosophical Sciences* (trans. Wallace, W.). Oxford: Clarendon Press.

Hegel, G.W.F. (1977 [1807]) *Phenomenology of Spirit* (trans. Miller, A.V.). Oxford: Oxford University Press.

Hegel, G.W.F. (1996 [1840]) *Lectures on the History of Philosophy* (trans. Haldane, E.S. and Simson, F.). Atlantic Heights, NJ: Humanities Press.

Heidegger, M. (1962 [1927]) *Being and Time* (trans. Macquarrie, J. and Robinson, E.). Oxford: Blackwell.

Henderson, G. (2013) *Value in Marx: The Persistence of Value in a More-Than-Capitalist World*. Minneapolis, MN: Minnesota.

Ilyenkov, E.V. (1977 [1974]) *Dialectical Logic: Essays on Its History and Theory* (trans. Campbell, J. and Creighton, H.). Moscow: Progress.

Inwood, M.J. (1992) *A Hegel Dictionary*. Oxford: Blackwell.

Jenkins, S. (2010) 'Hegel on Space: A Critique of Kant's Transcendental Philosophy'. *Inquiry*, 53(4), pp. 326–355.

Kant, I. (1992) 'Concerning the Ultimate Ground of the Differentiation of Directions in Space'. In Walford D. and Meerbote R. (eds) *The Cambridge Edition of the Works of Immanuel Kant: Theoretical Philosophy, 1755–1770*. Cambridge, MA: Cambridge University Press, pp. 365–372.

Kant, I. (1996 [1781]) *Critique of Pure Reason* (trans. Pluhar, W.S.). Indianapolis, IN: Hackett.

Krzhizhanovsky, S. (2006) *Memories of the Future* (trans. Turnbull, J.) New York: New York Review Books.

Lenin, V.I. (1976 [1914–1916]) 'Conspectus of Hegel's Book: The Science of Logic'. in *Collected Works, Volume 38: Philosophical Notebooks* (ed. Smith, S. and trans. Dutt, C.). Moscow: Progress.

Lyotard, J.F. (1998 [1984]) *The Assassination of Experience by Painting – Monory* (trans. Bowlby, R.). London: Black Dog.

Macherey, P. (2011 [1979]) *Hegel or Spinoza* (trans. Ruddick, S.M.). Minneapolis, MN: Minnesota.

Malabou. C. (2000) 'The Future of Hegel: Plasticity, Temporality, Dialectic', *Hypatia*, 15(4), pp. 96–220.

Marx, K. (1954 [1887]) *Capital: A Critique of Political Economy* (Vol I). (trans. Moore, S. and Aveling, L.). London: Lawrence and Wishart.

Marx, K. (1977 [1868]) 'Letter to Kugelmann, 11 July, 1868'. in McLellan, D. (ed) *Karl Marx: Selected Writings*. Oxford: Oxford University Press, pp. 524–525.

Marx, K. (2000 [1861–1863]) *Theories of Surplus Value, Books I, II and III* (ed. Ryazanskaya, S. and trans. Simpson, R.). New York: Prometheus.

Pilling, G. (1980) *Marx's Capital: Philosophy and Political Economy*, London: Routledge.

Saussure, F. (1959 [1916]) *Course in General Linguistics* (ed. Meisel, P. and Saussy, H. and trans. Baskin, W.). London: Peter Owen.

Simmel, G. (1971 [1907]) 'Exchange'. in Levine DN (ed.) *Georg Simmel: On Individuality and Social Forms*. Chicago, IL: Chicago University Press, pp. 43–69.

Simmel, G. (1978 [1907]) *The Philosophy of Money* (trans. Bottomore, T. and Frisby, D.). London: Routledge.

Sprinker, M. (ed.) (1999) *Ghostly Demarcations: A Symposium on Jacques Derrida's Specters of Marx*. London: Verso.

Stanek, Ł. (2008). 'Space as Concrete Abstraction: Hegel, Marx, and Modern Urbanism in Henri Lefebvre'. in Goonewardena K, Kipfer S, Milgrom R and Schmid C (eds) *Space, Difference, Everyday Life: Reading Henri Lefebvre*. London: Routledge, pp. 62–79.

Valéry, P. (1977 [1919]) 'The Crisis of the Mind'. in Lawler, J R. (ed) *Paul Valéry: An Anthology*. London: Routledge, pp. 94–107.

3 Locating heritage value

Kalliopi Fouseki, Joel Taylor,
Margarita Díaz-Andreu,
Sjoerd van der Linde and
Ana Pereira-Roders

Introduction

This chapter is based on work that took place as part of the European Network on
Heritage Values (H@V), a project funded by the JPI-JPHE Pilot Programming on
Cultural Heritage and Global Change. One of the central aims of this project is to
unfold the transnational and interdisciplinary meanings of the rather ambiguous
but extensively used concept of heritage value.

Within the framework of this project, the chapter aims to explore where heritage value resides in the heritage management field. It will do this by mapping
the epistemological geographies of heritage value – that is, how heritage value
is understood, defined and used across different disciplines. The chapter will not
focus on how other stakeholders, such as policymakers, the wider public and heritage visitors, understand the term "heritage value" since this has been explored
by recent research (e.g. Fouseki and Sakka, 2013) and since there is still scope for
further developments in this area. The chapter argues that heritage value constitutes an ambiguous concept in heritage management and that this ambiguity may
lead to misunderstandings, miscommunications and, consequently, mismanagement. Reser and Bentrupperbäumer (2005) have illustrated the consequences of
the ambiguity of the term "value" in the context of natural heritage sites. A similar
study does not exist for cultural heritage places. The chapter is directed towards
this gap in the literature by aiming to understand how "heritage value", a term that
is largely used by heritage professionals and academics, is discussed in different
disciplines.

The term "heritage value" (as well as the term "significance") became central
in heritage management discourses in the 1970s. However, reports produced by
the Getty Conservation Institute provided the foundation for a growing literature
on this matter (e.g. see de la Torre and Throsby, 2002; de la Torre, 2005). The
terms "value" and "significance" – often used interchangeably – dominate policy
(e.g. the Faro Convention on the Value of Cultural Heritage for Society, 2005) and
instrumental documents and charters (e.g. the influential Burra Charter, 1999). In
the UK, the organisations English Heritage and Historic England have adopted the
term and have further developed a typology of values, including community, historic, evidential and aesthetic (English Heritage, 2008). Similarly, the Department

of Culture, Media and Sports (DCMS) currently use the term mainly to connote the economic valuation of heritage participation (O'Brien, 2010). The terms "significance" and "cultural value" are commonly employed in the discourse of the Heritage Lottery Fund (Hewison and Holden, 2004). While the chapter will aim to locate value across disciplines, a next step should be to locate value across sectors and examine the gap between academic, policy and professional discourses.

First, we examine where heritage value resides in the disciplines of environmental studies and resource management; heritage studies and heritage management/conservation; and cultural economics. These disciplines were chosen not only because they relate to heritage but also because they have influenced each other in relation to heritage research. We proceed by elaborating on the findings of a brief literature review with preliminary research data derived from an anonymous, online survey that was conducted as part of the H@V project. This examined key questions and requirements of heritage professionals and academics on heritage value.

The chapter unveils a diversity of meanings that often leads to confusion, miscommunication and mismanagement of heritage (Reser and Bentrupperbäumer, 2005, p. 125) and to anger and frustration among members of interdisciplinary teams (Dillon et al., 2014; Bell et al., 2014). The ambiguity of the concept of heritage value and its impact on the management of heritage can be partly explained by the dominant influence of disciplines such as management science and environmental studies and by the distinct boundaries they tend to create (Becher and Trowler, 2001). Another source of ambiguity could be the working ethos of heritage management agencies such as museums, special heritage agencies in local authorities and governmental authorities. The language adopted by such agencies inevitably influences the development of heritage legislation, codes of ethics and other documents that underpin the management of heritage (e.g. see Smith, 2006).

In the case of heritage value, there are three levels of conceptual complexity that need to be considered. The first relates to the term "heritage" itself (see Harrison, 2013) which is often viewed as a social construct used interchangeably with "culture" and "tradition" although there is no agreed definition of heritage (e.g. Lowenthal, 1985). The second level relates to the term "value", which can refer "to fundamentally different phenomena, from individual human emotional response or judgment to shared convictions of how things should be, to a reading or calibration on a measuring instrument or scale" (Reser and Bentrupperbäumer, 2005, p. 27). The third level of conceptual complexity derives from the combined use of the terms "heritage" and "value". Indeed, by locating heritage in front of value, the meanings and references of the term "value" will change. It is this third level that the chapter will focus on.

Where does heritage value reside epistemologically and in heritage practice?

Framing the consideration of value as an issue that needs to be located is in itself problematic since this implies that value is a tangible entity that "can indeed have a locus" (Reser and Bentrupperbäumer, 2005, p. 140). This fixity of value is in contrast to the growing emphasis on value as a fluid, dynamic process and socio-political construct (Gibson and Pendlebury, 2009). However, despite the draw-back of adopting such a *locational approach*, this question can "force a critical and reflective consideration of current uses and meanings" and can also "address the apparent reality" according to which heritage values are "ostensibly and rou-tinely identified, located and measured impacted and protected" (Reser and Ben-trupperbäumer, 2005, p. 140).

An examination of how heritage value is discussed in cultural economics, environmental studies and heritage literature can reveal the multiplicity of the meanings with which the concept is interlinked. In cultural economics, heritage is viewed as a public good (e.g. see Sable and Kling, 2001), and thus, value is directly referring to the benefits to the public.

Often value is subdivided into private, market, social and non-market value (including aesthetic, cultural, option, bequest and existence) and thus is often associated with public benefit (Sable and Kling, 2001). Another common variant is use value and non-use value (i.e. option, bequest, existence) (Throsby, 2010). The assessment of value in cultural economics is synonymous with valuation and numerical measuring of the use or non-use, market and non-market economic and cultural value. In this field, value is assigned a numerical value, classified into distinct types or listed as criteria of significance (Throsby, 2010). The preferred term is "cultural" rather than "heritage" value, which is separated from the purely economic values.

In the literature on environmental conservation and history, heritage value is defined as the natural history, information storage and habitat for rare, archaeo-logical uses and current human uses (Smardon, 2006). Another interesting distinc-tion is that of "held" and "assigned" value (Seymour et al., 2010), which mirrors a long-running philosophical debate on the existence of "intrinsic" value (Zim-merman, 2010). Value in environmental studies in recent years tends to focus on "human value" and thus emphasis is on value as a social perspective, an ecological perspective, a psychological perspective and so on (Seymour et al., 2010). Other scholars in the field tend to understand value as a sociocultural and economic ben-efit derived from heritage, and they have been influenced by cultural economists (e.g. Alberini and London, 2009).

In the heritage studies literature is a gradual transition from the initially inter-changeable use of cultural and heritage value (Powell, 2000) and the division between intrinsic and extrinsic value (Carter and Bramley, 2002) to the adoption of value as intangible qualities that question the usefulness of compiling lists of criteria of significance (Smith, 2015). Interestingly, in archaeological site man-agement, values acquire tangible dimensions and are often classified into various types of values (such as economic, cultural, aesthetic, etc.) that can be measured (Avrami et al., 2000). Value is also increasingly defined around narratives (Walter, 2014). In urban heritage, the distinction between tangible and intangible value seems to remain relevant (McClelland et al., 2013).

Thus, value means different "things" to different professional groups, depending largely on the discipline represented but also on the wider context in which "value" is deployed. The recent UK Arts and Humanities Research Council project on cultural value is an explicit attempt to examine and then advocate for the benefits of culture in ways that include but are also distinct from the economic. This call also reflects governmental requirements from organisations such as English Heritage to provide evidence for the non-economic "value" of heritage and participation in heritage, emphasising impact of heritage participation on well-being. Thus, value can be understood as benefits, impact, outputs, outcomes, meanings, significance, narratives or all of these at the same time. With such a plurality of meanings, it is inevitable that misunderstandings and miscommunications occur. We now turn to examine the plurality of heritage value as understood by heritage professionals from different countries on the basis of data collected from the online anonymous survey of the H@V project.

Methodology

An online anonymous survey was carried out as part of the H@V project to gain an overview of how heritage professionals from different disciplines and countries conceptualise heritage value. It also examined their attitudes towards various qualitative or quantitative value typologies. This short survey informed the design of an in-depth, more widely circulated survey currently in progress. Due to limited available resources and the geographical disparity of the target audience, an online survey was viewed as the most suitable method. Given previous experience, during which we conducted a survey with experts (Dillon et al., 2014), we expected a high response rate.

The questionnaire comprised open questions and closed questions, with optional space for elaboration. The length of the survey was determined on the basis of the limited time that heritage professionals usually have to fill in surveys (Lauer et al., 2013, p. 338). The survey was available in Spanish, Catalonian, Greek and English. This chapter will focus on the analysis of the findings derived from the English survey, for which 108 responses were received.

The questionnaire was distributed to relevant professional networks and academic disciplines, including museums, heritage sites, heritage management, conservators, archaeologists (including respondents from the International Council of Museums-Committee for Conservation, the International Council on Monuments and Sites, the Association for Critical Heritage Studies and other relevant email lists). One of the main drawbacks of the survey is that the sampling of the respondents is random, so it is impossible to generate a representative sample of all the stakeholders (Bethlehem, 2009). Thus, the conclusions drawn in this chapter will apply only to the "frame population" – that is, the self-selected respondents who were contacted and responded – rather than to the full target population (Bethlehem, 2009, p. 27). Although a self-selection survey is limited, for the purposes of this particular project, such a survey, relying on a frame population, best suited the aims of this project and the available resources.

Hermeneutical framework

The data were analysed following the classification of disciplines proposed by Becher and Trowler in their seminal study *Academic Tribes and Territories* (2001). Through the analysis of 221 interviews with academics from a diverse range of disciplines,[1] their study showed that how "academics engage with their subject matter and the narratives they develop are important structural factors in the formulation of disciplinary cultures" (Becher and Trowler, 2001, p. 24). By "disciplinary cultures", they refer to the set of values, attitudes and ways of behaving that are articulated by a group of academics that represent a particular discipline (Becher and Trowler, 2001, p. 23).

Their sociological study informed the development of a classification of disciplines that was based on the cultural and social characteristics of each discipline. Despite the disadvantages of the proposed classification, as explained later, their classification system provides a useful tool for interpreting the online survey data. It is one of the first and most systematic studies in understanding the attributes of disciplines. Becher and Trowler are also the first to classify and group disciplines that share common characteristics. Thus, their work provides a useful tool for identifying trends and attitudinal patterns that can explain why certain professionals behave in certain ways. Because heritage management is interdisciplinary, an in-depth understanding of disciplinary cultures and attitudes can facilitate collaborative practices in the field.

Becher and Trowler classified the disciplines into four main groups:

1. soft applied
2. hard applied
3. soft pure
4. hard pure

But they point out that the boundaries between the hard/soft and pure/applied cannot be located with much precision (Becher and Trowler 2001, p. 39). Hard-pure disciplines refer to pure sciences (e.g. physics) and are concerned with universals, quantities and simplifications that result in discovery and explanation (Becher and Trowler, 2001, p. 36). Soft-pure disciplines refer to humanities (e.g. history) and pure social sciences (e.g. anthropology) and are concerned with particulars, qualities and complications that result in understanding and interpretation (ibid.). Hard-applied disciplines refer to technologies (e.g. mechanical engineering) and use both qualitative and quantitative approaches to develop products and techniques (ibid.). Finally, soft-applied disciplines refer to applied social science (e.g. education, law) and are concerned with enhancing professional practice through the use of case studies, resulting in protocols/procedures (Becher and Trowler, 2001, p. 39).

The classification of disciplines into hard pure, hard applied, soft pure, soft applied is based mainly on the "cognitive characteristics" of each discipline – for instance, the methods they use, questions they pose and theories they propose. In addition to the "cognitive classification", the authors attempted to classify disciplines into convergent and divergent social characteristics. Convergent disciplines tend to maintain uniform standards and procedures, whereas divergent disciplines tend to tolerate "a greater measure of intellectual deviance" (Becher and Trowler, 2001, p. 185).

This approach comes with certain limitations. Although the adoption of this classificatory system facilitates drawing general conclusions, there is a risk that each discipline is presented as a homogenous entity overlooking differences in each discipline. Indeed, several responses to the questionnaire highlight this diversity. Despite the risk of oversimplification, general tendencies can be drawn, which can provide an overall review of disciplinary attitudes towards heritage value.

Survey findings

Approaches to heritage value

A diverse range of disciplines were represented among the respondents (Table 3.1), which is indicative of the interdisciplinary nature of heritage studies, conservation and management. This interdisciplinary nature reinforces the need to unpack ambiguous concepts that are largely used in heritage practice and theory – such as the concept of heritage value. The disciplines were classified following Becher and Trowler's classification system and analysis, taking into account that the classificatory boundaries are not always clearly distinct (Table 3.1).

The respondents lived in different parts of the world, with the majority living in Northern Europe (especially the UK) (35%). Central Europe (17%), the US (14%)

Table 3.1 Disciplines represented in the survey classified following Becher and Trowler's classification and analysis

Discipline	Hard/Soft	Pure/Applied	Convergent/ Divergent	% of Respondents in the Survey
History	Soft	Pure	Convergent	3.8
Geography	In between	Pure	Divergent	2.8
Sociology	Soft	Pure or Applied	Divergent	2.8
Law	Soft	Applied	Intermediate	2.8
Economics	In between	Pure or Applied	Convergent	3.8
Anthropology	Soft	Pure	Convergent	7.8
Archaeology	Soft	Applied	Divergent	13.8
Architecture	Soft	Applied	Intermediate	14.2
Heritage Science	Hard	Pure	Divergent	5.8
Heritage Management	Soft	Applied	Divergent	5.8
Heritage Studies	Soft	Pure	Divergent	8.3
Museum Studies	Soft	Applied	Divergent	4.3
Education	Soft	Applied	Divergent	0.5
Ethnology	Soft	Pure	Intermediate	
Conservation	In between	Applied	Intermediate	16

and Southern Europe (14%) were also well represented. Fewer respondents came from Latin America (6%), Australia (5%), Southeast Asia (3%), Northern Africa (1%), the Middle East (1%) and India (1%).

The first question of the online questionnaire prompted the participants to provide their own definition of heritage value. The responses were coded into the following categories/approaches towards heritage value, including tradition, nostalgia/memories, intrinsic/intergeneration, benefits, human needs, narratives, types/typologies, thing, process, criteria and significance. Possibly not surprisingly, the majority of the respondents (19%) defined value as the significance or importance assigned to heritage that justifies its preservation. For one respondent, for instance, heritage value is "the key to why heritage is important to those who admire and live around it and why [the resource] needs to be protected and enhanced" (architectural conservation and planning, UK). For another, value is synonymous with "the significance placed on aspects of the past both tangible and intangible by communities which may or may not be recognised through legislation" (archaeologist, UK). Significance and value are thus used interchangeably as the justifier for preserving heritage.

The second most frequent response (15%) is closely related to the first and associates value with the "criteria" or "attributes" of significance or the whys that justify heritage preservation. As one participant put it, "for me, heritage value means the attributes that we as human beings associate with heritage that give heritage worth, meaning and importance and that make its understanding and preservation important" (heritage science and chemistry, UK). Another respondent defined value as a "set of chosen characteristics that are used to define what is worth protecting and preserving" or "the series of reasons why a specific heritage site or object is important to different stakeholders" (conservation science, Mexico). Of the respondents, 14% defined value as a dynamic sociocultural and political construct or process with particular emphasis on identity construction. For instance, a respondent noted that value is "what transforms ordinary objects, places, beliefs, cultural practices or past events in actual and real heritage" (archaeologist, Argentina). Another defined heritage value as "the aspirations we associate with heritage, such as the importance of heritage for identity, or the importance of heritage for social and political cohesion" (law, Brazil).

Of the respondents, 11% provided specific examples of heritage under their definition of value, such as "historical buildings, the built environment, rural landscapes (which have also been built), cultural foods, traditions and practices" (sociologist, Canada). Thus, their definition of heritage encompasses mainly the "what people value about heritage" more than the how and why. Of the respondents, 8% consciously or unconsciously defined value as distinct types or typologies. For instance, one respondent stated that heritage value is a "different kind of value (ethical, economic, social, political, etc.) attributed to a 'heritage object (material or immaterial)' by a group of actors (local community, academics from different disciplines, politician, etc.)" (political economist, France). For another, a holistic approach was taken by identifying value as "symbolic, historical, cultural, social, economical value", and yet another focused on "information", "aesthetics" and "economics".

Heritage value was defined by fewer respondents as narratives and meanings; human needs and morals; or tradition. One respondent, for instance, stated that heritage value is "a confusing term that I am not sure means the 'value' of material and/ or immaterial heritage, or human value passed on by people as part of their cultural heritage" (communication and cultural studies, UK). Along the same lines, another respondent defined value as "customs and beliefs that I have been brought up with that differ from what someone else has been brought up with" (discipline not identified). The intrinsic value or ethical obligation to transmit for future generations has also been emphasised by some respondents, and thus, value is the medium to achieve this transmission; for example, "value that [is] attached to tangible or intangible material held by a certain community that [is] passed on from generation to generation" (discipline not identified). For others, value is of a "tangible" nature and connotes the sociocultural or economic benefits that emerge from the protection of heritage. Thus, value is defined as the value "that heritage brings to and offers cultures, societies, communities and individuals in the broadest sense, rather than just in the economic sense" (museum studies, UK) or how "heritage can help us to create a better understanding of our common future" (conservation, Netherlands) and to contribute to "contemporary life and future development" (architecture-urbanism, Croatia). For a few, "heritage value is about feeling nostalgic and remembering or keeping connections to past cultural contexts" (heritage studies, Canada).

Finally, 7% of the respondents defined value as the narratives and meanings assigned by a wide range of groups, emphasising *who* values rather than on *what* and *why* is valued. For instance, a participant stressed that heritage value is

> what local people mean when they say, "this is our heritage". It depends on local value about what heritage constitutes. It has a wide range of meanings that can only be gleaned by asking local people what they think it means. However, it does include both intangible and tangible aspects.
>
> (anthropology, US)

In sum, it becomes obvious that heritage value is viewed in different ways by different professionals. This can be potentially problematic from a communication and collaboration point of view. For some heritage professionals and academics, heritage value constitutes a tangible concept, a "thing" or a "benefit" that can be described, measured, classified and assessed. For others, heritage value is mostly related to the why of heritage preservation and the how of its contribution to memory, identity and decision-making for present and future generations. A smaller percentage of respondents is concerned with the who of heritage value assignment. The following section will identify differences and correlations between disciplines and conceptualisations of heritage value.

Locating approaches to heritage value across disciplines

A cross-tabulation between the various disciplines and different approaches to heritage value revealed that applied disciplines are more likely to define value as a process or a sociopolitical construct, also highlighting the importance for establishing criteria or attributes of significance. Applied disciplines were also more likely to refer to value typologies and to define value as benefits (e.g. almost 90% of those from applied disciplines defined values as typologies, while only one respondent from pure disciplines used this term). On the contrary, pure disciplines tended to define value as meanings and narratives (90%). Thus, there is arguably a fundamental difference between the "tangible", describable, measurable nature of value adopted by applied disciplines and the "intangible", narrated nature of value adopted by pure disciplines.

A cross-tabulation of hard and soft disciplines with approaches to heritage value showed that hard disciplines were more likely to focus on value as "criteria of significance" or as "typologies" (50% of the respondents). Due to the small percentage of hard disciplines represented in the sample, the differences cannot be judged as statistically significant, but the tendencies observed are worth further exploration.

Overall, soft-applied disciplines tend to define value as criteria, benefits and significance and as a sociocultural or political process (more than 80% of the respondents under this type of discipline). Hard-applied disciplines are more likely to identify value as a series of typologies or criteria (50%). The emphasis of applied disciplines on these approaches can be explained by applied disciplines tending to be more concerned with practical implications and applied solutions to real-world problems. Hard-pure disciplines, on the other hand, were more inclined to define value as something tangible (20%), and the soft-pure disciplines identified heritage value as morals, memories and meanings (82%).

A closer look at some of the disciplines reveals that museum professionals emphasise value as benefits, while economists define value as specific types or typologies of value (Figure 3.1). The identification of value as significance or criteria/attributes of significance underpins mainly the disciplines of heritage science, conservation and heritage management, while heritage studies are divided between value as meanings and value as significance. Anthropologists are mostly referring to value as meanings or narratives, while for archaeologists and architects, value is either equal to significance or to a sociocultural and political process.

The current data do not reveal a correlation between attitudes to value typologies and approaches to heritage value. The only exception is that the respondents who defined value as morals or nostalgia and tradition tend to disagree with value typologies. Not surprisingly, the respondents who agree with value typologies tend to agree with quantitative methods for value assessment, but those who disagree with typologies do not necessarily disagree with quantitative methods. No statistically significant differences were observed between hard and soft disciplines and attitudes towards value typologies.

Hard disciplines tended overall to agree with the use of quantitative methods (25%) for assessing heritage value, while soft disciplines tended to disagree (82%). Overall, soft-pure and hard-pure disciplines were more likely to disagree with value typologies, while soft-applied and hard-applied disciplines were inclined to agree. Although variations occurred among disciplines in terms

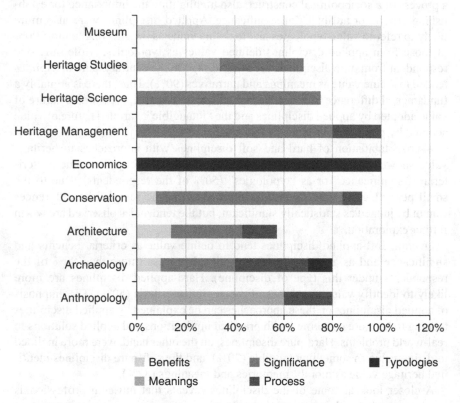

Figure 3.1 Approaches to heritage value cross-tabulated with specific disciplines

of attitudes towards the use of quantitative methods and value typologies, the response towards the use of qualitative methods was accepted by all disciplines, including hard-pure and hard-applied disciplines. This indicates that research and work associated with heritage is viewed largely as a qualitative subject.

Discussion

The analysis illustrated that the epistemological differences among various conceptualisations of heritage value occurring in the literature are also reflected in heritage practice. The survey unveiled 11 distinct approaches to heritage value:

1 tradition
2 nostalgia and memories
3 intrinsic and intergenerational
4 economic and sociocultural benefits
5 human needs/morals and beliefs
6 narratives and meanings

7 types and typologies
8 "thing"
9 process/construct
10 criteria/attributes of significance
11 significance

In other words, heritage value was defined by the survey participants in terms of *what* heritage is, *for what reason/why*, *by whom* and *for whom* heritage is preserved. Indeed, while for some respondents, heritage value is a tangible entity ("thing" or a "benefit") that can be described, measured, classified and assessed, others are mostly concerned about why heritage is preserved and how its preservation contributes to memory, identity and decision-making for present and future generations. A smaller percentage of respondents were concerned mainly with who assigns heritage value and its associated meanings.

This chapter explored the impact of particular disciplines in shaping certain attitudes and approaches to heritage value. Based on Becher and Trowler's (2001) assertion that academic disciplines form a distinct community with particular cultural and social characteristics, we examined the extent to which hard-pure, hard-applied, soft-pure and soft-applied disciplines differ in their approaches to heritage value. Wider factors, such as cultural context, country of origin, institutional ethos, personality of individuals, also contribute to the ambiguity of the concept of heritage value. Nevertheless, this chapter provides a starting point for understanding the disciplinary basis for those multiple layers of ambiguity around the term "heritage value".

For soft-pure disciplines, heritage value revolves more around who and what the public value as heritage. For soft-applied disciplines, emphasis was placed on heritage value as a list of criteria of significance or benefits that can justify decisions about preservation. In other words, it is the why that heritage is valued that really matters for these disciplines. Hard-applied disciplines adopt a more systematic (echoing scientific methods and techniques) approach to defining heritage value, emphasising the need for using existing or developing suitable value typologies. It is thus the how that heritage value is measured and assessed that matters. Hard-pure disciplines were not well represented in the sample of responses. Existing data show that they tend to define heritage value as a tangible entity that can be described, measured and assessed. Indeed, as Becher and Trowler (2001) have shown, in hard-sciences it is the methods that tend to determine the choice of the problems, whereas in soft disciplines it is the problems that determine the methods. Moreover, the fact that applied disciplines were more concerned about the whys and the hows, whereas the pure disciplines were more concerned about the what and who can possibly be explained by the fact that "pure knowledge, though increasingly vulnerable to epistemic drift, is essentially self-regulating, and applied knowledge, though occasionally prone to academic drift, is in its nature open to external influence" (Becher and Trowler, 2001, p. 185). As a result, both hard-applied and soft-applied disciplines are "amenable to outside intervention" (Becher and Trowler, 2001, p. 190).

Furthermore, the data reveal the complex and political nature of heritage value and, consequently, heritage management. As previously mentioned, approaches towards heritage value are driven by the who, what, why and how of value. These questions are fundamental in critically thinking about and then deciding on heritage value. As Taylor has shown (2014), the factors of who, what, when and how play a decisive role in "affecting choices and policies in a flexible way" (Taylor, 2014, p. 3): the factors do not stand in isolation; they interrelate. Two additional factors were revealed by the survey that could be incorporated into this model: the why factor and the for whom.

Going back to the initial question, where does heritage value reside epistemologically? The answer is that heritage value is located across a diverse range of disciplines imbued by different meanings that can partly be explained by the distinct culture and way of working of each discipline. The question that is more critical is how each discipline interprets and implements such an ambiguous concept in a critical manner that is of benefit for all those involved in heritage preservation.

Conclusion

"Heritage value" is unquestionably an ambiguous term that can potentially cause miscommunications in collaborative, interdisciplinary heritage projects, although this requires further exploration. The ambiguity of the term can be explained by the different cultural and ideological models, methods and approaches used by various disciplines and other factors (e.g. the institutional ethos of organisations, legislation and politics). Thus, an attempt to rename heritage value may not be successful. What is critical for avoiding miscommunication and mismanagement is to reach at the beginning of a project a consensus about what is meant by this term while adopting a flexible attitude towards the definition of the term. In the context of heritage, the term could be replaced by a phrase of heritage value as the what, why, how, by whom and for whom heritage is valued. By attributing the list of these questions, all involved in a heritage project will be required to reflect on the various elements included in this term.

This chapter begins an unfolding of the complexity and ambiguity of heritage value in order to tackle miscommunication. However, more research needs to be conducted. One area for further research is to look more closely, before looking at other disciplines, at heritage studies, heritage management, heritage science and heritage conservation. All these emerging fields advocate an interdisciplinarity and thus constitute divergent disciplinary areas open to dialogue and collaboration. However, each of those fields is informed or influenced by different core disciplines. Heritage science, for instance, is highly informed by hard-pure sciences. Heritage conservation is influenced mostly by hard-applied disciplines. Heritage management tends to be a divergent, soft-applied discipline that is highly informed by soft-applied disciplines (such as archaeology and architecture). Heritage studies, on the other hand, draw mainly on knowledge from soft-pure (e.g. anthropology, cultural studies) and soft-applied (e.g. sociology) disciplines. Indeed, an investigation of the extent to which the aforementioned emerging disciplines form distinct "academic tribes" with sociocultural characteristics and of their relationship with more traditional disciplines from which they have been influenced would allow a more in-depth understanding of how and where ambiguous concepts – such as that of heritage value – emerge, develop and affect practices and policies and would help us think more clearly about what can be done to mitigate the negative effects of such ambiguities.

An opportunity for further research is to look more closely at the policy discourses on heritage value and juxtapose these with academic and professional discourses and with how the wider public understands or assigns value to heritage. Finally, an area of research that is critical is to explore the impact (if any) of the ambiguity of this term in heritage practice.

Note

1 Biology, chemistry, economics, geography, history, law, mathematics, mechanical engineering, modern languages (French, German, Spanish and Italian), pharmacy, physics and sociology. Most of the aforementioned disciplines (including chemistry, economics, geography, history, mechanical engineering, physics and sociology) are included in the survey.

References

Alberini, A. and London, A. (2009) 'Valuing the Cultural Monuments of Armenia: Bayesian Updating of Prior Beliefs in Contingent Valuation', *Environment and Planning A*, 41(2), pp. 441–460.

Avrami, Erica C., Randall Mason and Marta De la Torre (2000) *Values and Heritage Conservation: Research Report*. Los Angeles, CA: Getty Conservation Institute. Available at: http://hdl.handle.net/10020/gci_pubs/values_heritage_research_report

Becher, T. and Trowler, P.R. (2001) *Academic Tribes and Territories* (2nd ed.). Buckingham and Philadelphia: The Society for Research into Higher Education & Open University Press.

Bell, N., Strlić, M., Fouseki, K., Laurenson, P., Thompson, A. and Dillon, K. (2014) *Mind the Gap: Rigour and Relevance in Collaborative Heritage Science Research*. Available at: www.nationalarchives.gov.uk/documents/mind-the-gap-report-jan-2014.pdf

Bethlehem, J.G. (2009) *Applied Survey Methods: A Statistical Perspective*. Hoboken, NJ: Wiley.

Carter, R.W. and Bramley, R. (2002) 'Defining Heritage Value and Significance for Improved Resource Management: An Application to Australian Tourism', *International Journal of Heritage Studies*, 8(3), pp. 175–199.

De la Torre, M. (ed) (2005) *Heritage Values in Site Management: Four Case Studies*. Los Angeles: Getty Conservation Institute.

De la Torre, M. and Throsby, D. (2002) *Assessing the Values of Cultural Heritage: Research Report*. Los Angeles: Getty Conservation Institute.

Dillon, K., Bell, N., Fouseki, K., Laurenson, P., Thompson, A. and Strlić, M. (2014) 'Mind the Gap: Rigour and Relevance in Collaborative Heritage Science Research', *Heritage Science*, 2(11), pp. 1–22.

English Heritage (2008) *Conservation Principles: Policies and Guidance for the Sustainable Management of the Historic Environment*. Available at: https://historicengland.org.uk/images-books/publications/conservation-principles-sustainable-management-historic-environment/

Fouseki, K. and Sakka, N. (2013) 'Valuing an Ancient Palaestra in the Centre of Athen: The Public, the Experts, and Aristotle', *Journal in Conservation and Management of Archaeological Sites*, 15(1), pp. 30–44.

Gibson, L. and Pendlebury, J. (eds) (2009) *Valuing Historic Environments*. Surrey: Ashgate.

Harrison, R. (2013) *Heritage: Critical Approaches*. London: Routledge.

Hewison, R. and Holden, J. (2004) *Challenge and Change: HLF and Cultural Value. A Report to the Heritage Lottery Fund*. DEMOS. Available at: www.hlf.org.uk/challenge-and-change

Lauer, C., McLeod, M. and Blythe, S. (2013) 'Online Survey Design and Development: A Janus-Faced Approach', *Written Communication*, 30(3), pp. 330–357.

Lowenthal, D. (1985) *The Past Is a Foreign Country*. Cambridge, MA: Cambridge University Press.

McClelland, A., Peel, D., Lerm Hayes, C-M. and Montgomery, I. (2013) 'A Value-based Approach to Heritage Planning: Raising Awareness of the Dark Side of Destruction and Conservation', *Town Planning Review*, 84(5), pp. 583–603.

O'Brien, D. (2010) *Measuring the Value of Culture: A Report to the Department for Culture, Media and Sport*. Department for Culture, Media and Sport. Available at: www.gov.uk/government/publications/measuring-the-value-of-culture-a-report-to-the-department-for-culture-media-and-sport

Powell, J. (2000) 'Expanding Horizons: Environmental and Cultural Value Within Natural Boundaries', *International Journal of Heritage Studies*, 6(1), pp. 49–65.

Reser, J.P. and Bentrupperbäumer, J.M. (2005) 'What and Where Are Environmental Values? Assessing the Impacts of Current Diversity of Use of 'Environmental' and 'World Heritage' Value', *Journal of Environmental Psychology*, 25, pp. 125–146.

Sable, K.A. and Kling, R.W. (2001) 'The Double Public Good: A Conceptual Framework for "Shared Experience" Value Associated With Heritage Conservation', *Journal of Cultural Economics*, 25(2), pp. 77–89.

Seymour, E., Curtis, A., Pannell, D., Allan, C. and Roberts, A. (2010) 'Understanding the Role of Assigned Value in Natural Resource Management', *Australasian Journal of Environmental Management*, 17(3), pp. 142–153.

Smardon, R.C. (2006) 'Heritage Value and Functions of Wetlands in Southern Mexico', *Landscape and Urban Planning*, 74(3–4), pp. 296–312.

Smith, A. (2015) 'World Heritage and Outstanding Universal Value in the Pacific Islands', *International Journal of Heritage Studies*, 21(2), pp. 177–190.

Smith, L. (2006) *The Uses of Heritage*. London: Routledge.

Taylor, J. (2014) 'Recontexualising the 'Conservation Versus Access' Debate'. *ICOM-CC 17th Triennial Conference: Theory and History Conservation*. Conference paper.

Walter, N. (2014) 'From Values to Narrative: A New Foundation for the Conservation of Historic Buildings', *International Journal of Heritage Studies*, 20(6), pp. 634–650.

Throsby, D. (2010) *The Economics of Cultural Policy*. Cambridge, MA: Cambridge University Press.

Zimmerman, M.J. (2010) 'Intrinsic Versus Extrinsic Value', *Stanford Encyclopaedia of Philosophy*. Available at: http://plato.stanford.edu/entries/value-intrinsic-extrinsic/.

4 Making values visible and real, but not necessarily monetised

Wendy Miller

Introduction: discerning values

Heightened concern exists worldwide about food security for mainly urbanised populations (e.g. Steel, 2008), whether due to continuing hunger and malnutrition, soil erosion, climate change and wildlife extinctions (Ceballos et al., 2017) or future hiatuses in trading arrangements. This chapter looks at the many values involved in agriculture and the food systems through which populations secure – or fail to secure – the food they need or want. The focus is on UK allotments and other local food initiatives based on research in Plymouth, England. It uses the capital-assets framework, as developed in the sustainable livelihoods approach, as a tool to incorporate non-monetised values. The framework is used to identify the values and impacts of these food-producing systems, on human, social, economic, cultural, political and natural capital-assets, and it considers the contingent or contextual factors that mediate these impacts.

Values, whether economic, human, social, political, cultural or environmental, determine the decisions taken in all policy arenas. However, the values that are fed into policymaking are predominantly monetary – for example, the familiar GDP (gross domestic product). Critiques of measures such as GDP as an indicator of what matters have been long-standing and far-reaching. Given the heightening awareness of environmental, social and cultural concerns, broader indicators are increasingly taken into account, although they may still be "reduced" to a monetary value in final analyses. The many methods developed to account for other values over the past three decades include economic calculations such as contingent valuation and existing geophysical, geopolitical and socioeconomic datasets to produce indicators such as the Human Development Index or the Gross Happiness Index, developed in Bhutan.

Since publishing the Millennium Ecosystem Assessment (MEA, 2005), with headlines that global ecosystems provide an estimated US$33 trillion in value to societies, increasing efforts have been devoted internationally to ecosystem services assessments (EAs), national natural capital accounts (see for example the UK National Ecosystem Assessment, and the UN System for Environmental-Economic Accounting). Evaluations of ecosystem services at project, national and global levels have gained significant funding from research agencies and governmental and non-governmental organisations (DEFRA, 2008). These all use both monetary and physical indicators to assess asset bases and flows (Carpentera et al., 2009; Haase et al., 2014) [see Carver and Crawford, this collection].

The ever-increasing number of heuristics that attempt to calculate values and to settle contested priorities, such as life cycle analysis (LCA), often require in-depth knowledge, specialist expertise and significant amounts of additional data gathering such as surveys (Heynen and Robbins, 2005). EAs are criticised for "comparing apples and pears": for example, soil erosion in tonnes per hectare compared to biodiversity in numbers of species may not help policymakers make their decisions. EAs also involve converting ecosystem values to monetised values on the basis of questionable assumptions and can lead to skewed and/or incomplete results (McBurney, 1990). Further, as indicated by the trenchant critique from Daly (2005), in *Economics for a Full World*, many analyses are at micro-level and meso-level and completely omit any consideration of macro-level impacts, as suggested by the system's perspective on planetary boundaries (Rockström et al., 2009). According to calculations, humanity has exceeded the capacities of the environment on which life depends on parameters including biodiversity loss and the nitrogen cycle, much of which can be attributed to agriculture. Eco-footprint assessments have gained ground because they provide an easily accessible single numeric indicator but are again critiqued on the basis of the calculations behind the numbers.

Concerns are raised about the "capture by the neo-liberal agenda" of the different valuation methods discussed earlier and the devaluing of future generations (Dempsey and Robertson 2012). To counter this, techniques such as social return on investment (SROI) were developed for community stakeholders in social projects. These incorporate considerations of health and environmental well-being through proxy statistics from already-existing data. The method involves identifying all outcomes and indicators relevant to all stakeholders and often assigning financial proxies. However, SROI and most of the other tools still lack dimensions of democracy and justice (Ostrom, 1999). More recently, the political value of an active citizenship and "participatory democracy" has been increasingly acknowledged (Michels and de Graaf, 2010). However, the parameters of autonomy and self-reliance, which are key factors in food security, have not yet been included in most policy assessments. There therefore remains the need to continue to explore ways to include non-monetised values in decision-making.

Recognising values in food systems

As with other policy arenas, monetary values have long formed the basis of policymaking in food and agricultural systems and have largely been taken as a proxy for the sector's well-being. However, the recognition of multifunctional agriculture is one attempt to take into account the different values involved in decision-making in the sector (e.g. see IAASTD, 2008). For example, environmental and social parameters have been incorporated through concepts such as the popularised food miles or fair trade (Pretty et al., 2005). An overview of the widening spheres of food system values recognised over the last seven decades is illustrated in Table 4.1 and shown in comparison to the "baseline" of household non-monetised food production as exemplified by the UK's allotment system – that is, land made available "at reasonable rent" by local authorities for households to grow food.[1]

Table 4.1 illustrates the ever-widening spheres of values and concerns relevant to food and agriculture systems. Attempts in the early 21st century to take into account these values have led to a bewildering array of indicators. An EU conference called "Where next for Sustainable Development" held in 2010 heard policymakers' urgent pleas for some form of agreement between all the different "toolmakers", to provide a clear and common basis on which to make decisions, in language that is accessible to non-specialists. While sustainable intensification remains the oft-stated policy goal for industrial agriculture, international organisations and social movements increasingly recognise the contributions of agroecological and small-scale, or self-provisioning, food production towards biodiversity and food security goals. Alongside international and national efforts to develop ecosystem services indicators (TEEB, 2018), a burgeoning academic literature has developed on community food initiatives, their impacts and contingent factors (e.g. McClintock, 2010). The following sections explore the diverse capital-assets framework that can be usefully applied to the different values involved in food systems, drawing on a case study in Plymouth, England, to illustrate their application.

Conceptualising values in food systems

This section looks at concepts developed to encompass more-than-economic spheres of concern and how they might help to resolve value conflicts between different priorities. It explores the idea of diverse economies formulated by Gibson-Graham (2003, 2008) to help make different values visible, followed by an overview of the capital-assets framework as a means of accounting for these values.

Diverse economies

Walker (2007), along with other critics (e.g. Dryzek, 1997), suggests that to effect change, counter-narratives to economic valuation need to be created. Gibson-Graham (2008) introduced the concept of "diverse economies", contending that the market economy is merely the tip of any iceberg and that non-monetised economies (e.g. housework) far outstrip the value of monetised economies. Diverse economies encompass acts of reciprocity, unpaid work and barter systems, as have long existed in traditional food economies.

Gibson-Graham (ibid.) suggests the unseating of monetised economies by "making visible the invisible" values. This might require the creation of new information in constructing and framing problems and in their measurement and

Table 4.1 Widening spheres of values and concern in food systems

Decade	Theme	Values recognised/impacted	UK allotments and community food initiatives	Illustrated by
1950s	Post–world war productivism	Maximising yields, increased chemical inputs, mechanisation, decreasing employment in agriculture, convenience foods	Reduced demand for allotments	The Food System (Tansey and Worsley, 1995)
1960s	Environmental awareness	Damage to ecosystems by agrochemicals. Concepts of integrated pest management. Growing demand for "natural wholefoods"	Recognition of value of household food production to family budgets and nutrition and of leisure time in the natural environment	Silent Spring (Carson, 1962), The Thorpe Report (MLNR, 1969)
1970s	Questioning technologies	Recognition of oligopolistic control of agrochemicals, soil erosion, problems of excess packaging	Increasing demand for allotments, waiting lists rose by an estimated 1,600 percent in the decade[1]	Small is Beautiful (Schumacher, 1973), Limits to Growth (Meadows et al., 1976), The Complete Book of Self-Sufficiency (Seymour, 1976)
1980s	From atomistic individuals to international communities	UK PM statement: "No such things as society" Social inequality in the UK rose more than in any other comparable country First multilateral environment agreement (Montreal Protocol) First EU agri-environment schemes Fall of the Berlin Wall	Ecovillage movement for community food provisioning Demand for UK allotments fell	Our Common Future (Bruntland, 1987)
1990s	Democracy and justice	International Convention on Biodiversity, Earth Summit (Rio de Janeiro) 1992 World Trade Organisation set up in 1995 Anti-globalisation protests in Seattle 1999 Issues of food miles and fair trade	International food sovereignty and land rights movements, Via Campesina formed in 1992	The Politics of the Earth (Dryzek, 1997)

2000s	Multidimensional concerns	Systems approach recognised Millennium Ecosystem Assessment (MEA, 2005) "Sustainable Intensification" policy goal for agriculture Increased global demand for meat and dairy products	New urban community food initiatives, farmers' markets, veg. box schemes, community-supported agriculture Waiting time for a UK allotment up to 40 years in one city area[2]	*Hungry City* (Steel, 2008), *Feeding People is Easy* (Tudge, 2007)
2010s	Valuing the future	Heightened concerns for biodiversity loss ("The Age of Extinction" [Ceballos et al., 2017]), soil erosion and climate change Supermarket initiatives to redistribute "waste" food	Food banks in high demand International and local umbrella networks of food-related social movements, e.g. Food Charters, UK Landworkers Alliance, Extinction Rebellion	Online and e-campaigns

Source: Author

1 Crouch and Ward (1997)
2 www.theguardian.com/news/datablog/2011/nov/10/allotments-rents-waiting-list [accessed 29 November 2018]

proposed solutions. However, the need for new information may be met by reanalysing existing data in new ways. Making the invisible visible for any initiative, programme or policy can be in (a) descriptive terms of outcomes; (b) in non-comparable indicators, such as reduced obesity and reduced days of staff absence; and/or (c) in financial terms, according to available data, such as cost of obesity to the health service and cost of replacement staff. The key issue is that all outcomes of interest to all stakeholders are included and are made visible.

Capital-assets framework

It is evident that traditional monetary valuations do not capture everything that really matters to people, organisations and ecosystems. Bourdieu (1986), while suggesting that economic capital lay behind all other forms of capital, also contended that other forms of capital need to be accounted for:

> It is in fact impossible to account for the structure and functioning of the social world unless one reintroduces capital in all its forms and not solely in the one form recognized by economic theory.
>
> (Bourdieu, 1986, p. 241)

A "non-hierarchical" system's perspective can be gained through the lens of the capital-assets framework (CAF). Initially developed within international development studies, CAF provides a means of capturing the dimensions identified as key components for sustainable livelihoods (Scoones, 2009). Many iterations of CAF exist (see e.g. TEEB, 2018); Figure 4.1 sets out the scheme drawn on here.

The capital-assets framework is now employed for several community development initiatives in the US as the community capitals framework (Emery and

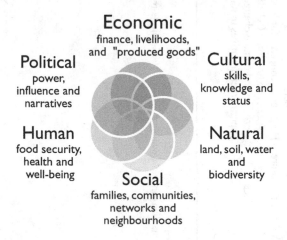

Figure 4.1 The capital-assets model

Source: Author adapted from various

Flora, 2006). Although monetary proxies are ubiquitous and often useful, there are other key indicators of the different dimensions (human, social, cultural, natural, economic and political) that are also meaningful to policymakers. For example, on health issues (human capital), the percentage of population who are obese carries a strong significance for policymakers on which to base decisions, even if also converted to monetary cost. For social capital, levels of social isolation can be used, as measured through existing local government surveys on city populations. Similarly, for political capital, numbers contacting a public official or politician are possible measurements available at local, national and international levels (Pattie et al., 2004). For natural capital, if the concern is biodiversity, then levels of landspare or landshare can be determined without high levels of specialist expertise.[2] The next section gives an overview of a specific case in which these dimensions are considered.

Grounding values: diverse food networks in Plymouth, England

The capital-assets framework can be applied at any scale and in any policy arena. The example used to illustrate its application here draws on empirical research carried out between 2010 and 2013 as part of a doctoral programme on allotments and diverse food networks in Plymouth, England.

The city has a population of around 250,000 and is surrounded by an agricultural hinterland and the Atlantic Ocean to the south. South West England is a region with a low-wage economy, and employment levels in Plymouth largely depend on the public sector (health, "defence", and education). Research methods included participatory observation, interviews and reanalysing existing datasets and archival records. The primary focus was on allotments within the city as a "benchmark" against which to analyse other, newer, local food initiatives in the city and region.

Plymouth allotments

Allotments are the longest-standing form of local food production aside from domestic production on owner-occupied gardens and property. Since the inception of the UK allotment system as defined by Parliamentary Acts in 1845 and 1907 (see Miller, 2013 for detail), their recognised value has varied over time and has depended on wider contexts. During wartimes, allotments were valued as a source of survival food supplies and during the 1960s as a leisure opportunity. Since the financial crash of 2008, demand for allotments has risen all around the UK. There are around 32 sites with roughly one thousand individual plots in Plymouth, and in 2013 the waiting list for a plot was also around a thousand.

City community food initiatives

In the past two decades, the number of city community urban food initiatives has increased significantly, often with grant funding from local, regional or national funds. In Plymouth, Grow Efford was a project in the national regeneration programme Building Communities Initiatives, with full-time and several part-time project workers. Similarly, Diggin It, with National Lottery funding, employed two or more project workers and set up a professional growing area equivalent to ten individual plots on a city allotment site. This project provided volunteering opportunities as well as facilities for adult social services across the city. A school outreach initiative was developed, but the project's workers were lost as core funding dwindled.

Local and regional food initiatives

In peri-urban or rural areas, a plethora of enterprises, such as local farm shops, vegetable box schemes and community-supported agriculture have developed. Residents in Plymouth can choose between several vegetable box providers, including Riverford Farm, whose base is to the east of the city. Private, commercial farm shops on the outskirts of the city include Pick Your Own farms, used by city residents on a weekend trip out to the countryside. Another initiative is Tamar Grow Local (TGL), reliant on grant funding for paid workers who facilitate a range of projects that include a community-supported orchard and a honey cooperative. A key aim of TGL is to increase the number of land-based income-earning opportunities in the Tamar Valley, previously a prime horticultural area but now a place that has seen a steep decline in production in the face of increasing imports of fresh produce from countries with lower labour costs and more favourable climatic conditions.

City charter

From 2010 onwards, a collaboration of food-related projects, organisations and interested individuals, including academics, was given a small amount of funding by the city council, the University of Plymouth and the national Soil Association. Despite an unsuccessful application for lottery funding, it adopted the name Food Plymouth and launched a food charter, drawing on the pioneering work in Toronto, Canada. The group subsequently achieved funding from the EU programme DEAL in collaboration with French partners. The Food Plymouth alliance contributed to the rise in salience of food as a policy area within the new core strategy for the city council. The Plymouth Food Charter encompassed all the different dimensions of values outlined in the capital-assets framework mentioned earlier. A frequent contention expressed during Food Plymouth meetings was the need to communicate to different audiences: if speaking to an educationalist, then talk in terms of increased educational attainment, and if to health practitioners, then in terms of measures of health and well-being.

These brief descriptions of just some of the different food initiatives in Plymouth illustrate the different stakeholders and value priorities in the food system in one city region: household food provisioning (allotments), community participation, income-generation or enhanced policy awareness.

Operationalising values

The following section operationalises the monetised and non-monetised values outlined earlier, for Plymouth food initiatives, in a way that can be incorporated in decision-making. This application of the capital-assets framework accounts for both processes and outcomes (Perrons and Skyers, 2003), and it assesses the baseline (system "initial starting conditions") and contingent factors of any situation. The section concludes by suggesting how these techniques can be applied in other policy arenas and at other scales (regional, national and international).

Human capital

As shown in Figure 4.1, human capital denotes the health and well-being of individuals. Concerns around quality of food are expressed by many involved in local food initiatives, including allotment plot holders, alongside acknowledgement of the benefits of the exercise involved and being in the natural environment. As one participant stated,

> "What better way to spend a few hours relaxing than gardening; there's always something new, and always something interesting. I can just do my own thin, and listen to the birds, fiddle about, sit down a bit, then go and do a bit of digging."

Not all allotmenteers are primarily concerned with increasing household food production, especially given the widespread availability of food through supermarkets. They may instead value "the restorative power of the natural environment" to reduce stress levels. However, notably, since the financial crunch of 2008 and ensuing "austerity measures", food security has risen up the political agenda: increasing numbers of city residents seek basic supplies from charitable food banks and/or are on the waiting list for an allotment plot (Miller, 2013).

The proportion of fresh food in diets can be used as one of the possible indicators for human capital. Taken at its simplest, and using readily available data, the proportion of the population meeting their recommended 5 portions of fruit and vegetables a day ("5-a-day") can be extrapolated from national statistics, applied to Plymouth demographics and supplemented by qualitative research to add depth and the "personal interest" that helps issues rise up policy agendas (see Table 4.2). For example, research for the World Cancer Research Fund has shown that only 17% of social groups C2, D and E eat their 5 a day, compared to 27% of groups C1 and AB.

Table 4.2 illustrates that non-monetised or community food initiatives that increase fresh food consumption have the potential to benefit human capital for 147,138 adults – as well as children – across the city. Two allotment tenants expressed how they valued increasing their availability of vegetables:

> "Runner beans don't freeze brilliantly, but they still give a taste of summer in the winter; I've still got beans from last year, and that's with giving loads away to the family and friends".

Table 4.2 Plymouth residents' intake of recommended 5-a-day

	Count (a)	%	numbers eating 5 a day (b)	numbers that could benefit from greater availability of fresh food in the city (a-b)
All people aged 16 and over in households (people)	187,417			
AB: Higher and intermediate managerial/administrative/ professional (people)	30,818	16	8,320	22,498
C1: Supervisory, clerical, junior managerial/administrative/ professional (people)	53,377	28	14,411	38,966
C2: Skilled manual workers (people)	33,905	18		
D: Semi-skilled and unskilled manual workers (people)	37,509	20	17,548	85,674
E: On state benefit, unemployed, lowest grade workers (people)	31,808	17		
Total				147,138

Source: ONS, Census 2001 data for Plymouth, World Cancer Research Fund, YouGov poll 2012

"You can tray-freeze courgette slices then bag them up, then you've always got something to add into stews, soups or whatever all through the year".

This supports findings by Alaimo et al. (2008), indicating that city food growers increase their intake of fruit and vegetables. It is suggested that discussions using numbers of people potentially affected alongside "vignettes" may have more policy salience than a monetised value of the cost of not meeting dietary recommendations. Analyses can be taken as far as required (and resourced) for any policy decision – for example, levels of exercise or time in the natural environment.

Economic capital

Economic capital is the most clearly acknowledged dimension in any policy discussion. The initiatives in this study range in economic impact and include saving money on household food budgets, time committed to voluntary activities (such as growing and gifting food, teaching/learning culinary skills) and income-generation or employment. The Campaign for Real Farming (CRF) developed a scenario suggesting that an estimated 10% of the adult population would be needed to work in the food production sector in order to provide "good food for everyone forever" (Tudge, 2011). For Plymouth, this would mean employing around 12,000 people, equivalent to the number of people currently unemployed in the city. Potential impacts of food projects on economic capital could also be determined by numbers desiring to grow their own food. The national Taylor Review (2008) reported that 33% of the population would like to do so, equating to over 60,000 adults in Plymouth who could reduce household food budgets in this way if given the opportunities. Instead of using a single monetary value, a fuller picture and the wider context can be given through readily available statistics and scenarios developed by policy and civil society organisations. There are of course many contingent factors for the CRF scenario to be realised, not least the amount of land allocated to food initiatives.

Social capital

The extent of social capital, in terms of networks and of community interaction, can be assessed by numbers participating in community events. In this instance, numbers of local authority social service "clients" benefiting from food projects can provide an indicator for use in decision-making. For example, an SROI evaluation of the lottery-funded programme Local Food in Fife (Lancaster and Durie, 2008) found that £1,687,441 of investment in local food projects had led to benefits valued at £11,756,563: a ratio of benefit to investment of nearly 7:1. It was estimated that the programme produced around 62% of benefits in the area of health, 26% in the area of community vibrancy and 8% in the area of education and skills.

If we apply this to our case study, using the Plymouth City Council budgets for 2013, if just 1% of the adult social care and the learning and family support budget were spent on food projects in the city, that could amount to £900,000 invested for a social return of £6,228,000, of which £1,619,280 could be estimated to contribute to reduced social isolation and to other capital dimensions (e.g. human and cultural). Such calculations could give weight to any stated need for allotment and community gardening opportunities to be made available to the city population.

Cultural capital

The cultural capital of education and skills gained through food projects include cooking, growing and eating and can be assessed by the numbers involved. Evaluations carried out for funded projects in the city have demonstrated impacts on skills through food projects, leading to employment in several instances. For example, a pilot Food Cultures project in Plymouth involving young men and older adults found that their 42 participants had all learned new growing and cooking skills, evidenced by short video clips. In the realm of school education, an evaluation of the national Food for Life programme (FfLP), which involves school children learning to grow and cook food through different aspects of curricula, showed increased attendance and improved learning outcomes (FfLP, 2011). This learning is also valued among allotment tenants:

"The blackbirds tell you when the sweetcorn is nearly ready . . . start pecking it open . . . then you just need to get to it before they do".

"It's like you're getting free lessons from real experts".

Thus, food-related skills and learning can be gained through project participation, formal education or allotment communities, enhancing cultural capital. These can be evidenced through project evaluations and qualitative studies and can also be quantified by extrapolating from existing studies, such as those on school attendance, to demonstrate impacts and contingent factors (such as project funding) for these to be realised.

Natural capital

A key motivation for involvement in local food initiatives is concern for the environment, but assessing ecological impacts involves intricate, contested estimates (see Pretty et al., 2005 for example) and often complex econometrics. Settling these differences requires specialist input, such as habitat surveys. Rather than attempt to account for all possible variables, the indicator of primary concern, whether biodiversity, carbon emissions or soil erosion, can be foregrounded. For example, a commitment to the Biodiversity Action Plan for Devon requires wildlife habitat provision, and the extent of landshare or landspare can be a simple metric applied to land for food production. In this case, rather than requiring input from expert ecologists, an assessment can be made from observations and maps correlated to existing land use and biodiversity data. Maps and photographic evidence (Figure 4.2) can combine to give policymakers a sense on which to base

Figure 4.2 Room for wildlife in Plymouth food gardens – bug hotel in a school garden
Source: Author

decisions. For quantitative measures, the environmental impacts of different food initiatives can draw on further evidence, such as ecosystem services indicators from other studies or Natural Capital Accounts.[3]

Political capital

In the UK, fast-increasing costs of welfare payments led to intense parliamentary debates in the 1880s about the need for "allocation of land to rent at reasonable rates" in the form of allotments, for reasons of self-reliance and autonomy. In what were dubbed the Allotment Elections for local authorities in 1889, the loss by Gladstone's government was attributed to not enough land being made available to "the poor" (Way, 2008; Boyle, 2012). Since then, land has been increasingly subject to monetary speculation. Urban land prices preclude allocations for non-monetised food production as a viable strategy for cash-strapped local authorities: the Plymouth Parks Department find they continually need to justify land assigned to allotments. Yet as illustrated by social movements such as MST, *Movimento dos Trabalhadores Sem Terra*, the landless workers movement in Brazil, and in many other locations and times globally, land redistributions are possible (Caldeira, 2008). Through comparing land allocations to food-growing initiatives to allocations for other policies, decisions over city resources can at first be made more visible, as in Table 4.3.

When considering political capital, numbers contacting officials can give some indication of "active citizenship". In terms of food supplies and self-reliance for households, a more meaningful indicator is the level of land allocated:. the level of resources leveraged into the policy area. Basing decisions on monetised values of urban land will clearly lead to low allocations for non-monetised food initiatives. Yet by bringing the different capital assets together, the multidimensional

Table 4.3 Land allocations in Plymouth in terms of allotment plot equivalent and land area

	No. of plots equivalent	Land area (ha)
New land brought into allotment cultivation 2010	39	1
Allotment land requirement for waiting list (half plot size 125 m²)	500	12
Current allotment provision (2012)	1,478	23
Allotment land requirement for waiting list (full plot size 250 m²)	1,000	25
Employment land requirement identified in Core Strategy (PCC, 2007)	2,480	62
Parks Department managed greenspace	38,000	950
Domestic gardens	70,560	1,764
Plymouth land area of greenspace (except domestic gardens) – 42%	141,120	3,528
Nat Ag Labourer's Union campaign 1885, 0.17 ha per person	160,000	4,000
Plymouth land area	336,000	7,930

Source: Plymouth City Council Employment Land Review [2006] and allotment statistics

values can be made visible at this point and fed into resource-allocation systems at any scale. Realising the potential impacts on health, learning, communities, economies and biodiversity may then enable greater numbers of decision-makers to view food-related initiatives as a valid policy option that can help to enhance all forms of capital-assets for urban populations – including balancing house-hold, local authority and national budgets. While the allotment system was foregrounded earlier, other food initiatives can also be assessed on the different dimensions, discussed further elsewhere (Miller, 2013).

Conclusion

The utility of the capital-assets framework in making multidimensional values visible is that each indicator or value can be foregrounded in turn according to interest, policy audience or salience. Further, the earlier illustration of the frame-work's application demonstrates the inter-linkages between the different capital assets and so provides the necessary system-level perspective (TEEB, 2018).

The capital-assets framework demonstrates that common languages between decision-makers in different arenas can be attained in discussions of values but perhaps also that it is a question of becoming multilingual: a basic literacy is required in all the key concepts from across both the social and natural arenas of inquiry. However, these need not be specialist or "high-level" skills, as suggested by the simple indicator of landspare or landshare, accessible to any observer.

As shown by this one case, non-monetised values can be made visible (Gibson-Graham, 2008) by using readily available proxy data that decision-makers can relate to, such as unemployment rates, health statistics and land allocations. As illustrated, these can be quantified and qualified to convey more information than a simple traf-fic light scheme on how the different capital-assets will be affected by any policy decisions. The capital-assets framework enables all stakeholders involved to locate values involved in food-related activities or any other field of interest. It seems that perhaps apples and pears can be meaningfully compared, on many dimensions.

Notes

1 This approach has been fruitfully used by Cook et al. (2010) in the areas of sustainability and teacher education.
2 Landshare involves ecological methods of production being used, and landspare is as conceived in EU policy, whereby areas are set aside for non-crop species. Agroecologi-cal land management practices, where food production and biodiversity both flourish, would obviate the need for this distinction, although assessment would again rely on specialist habitat surveys.
3 www.ons.gov.uk/economy/environmentalaccounts/bulletins/uknaturalcapital/ecosystem accountsforurbanareas

References

Alaimo, K., Packnett, E., Miles, R.A. and Kruger, D.J. (2008) 'Fruit and Vegetable Intake Among Urban Community Gardeners', *Journal of Nutrition and Education Behavior* 40(2), pp. 94–101.

Bourdieu, P. (1986) 'The Forms of Capital', in Richardson, J. (ed) *Handbook of Theory and Research for the Sociology of Education*. Westport, CT: Greenwood, pp. 241–258.

Boyle, D. (2012) *On the Eighth Day God Created Allotments*. London: Real Press.

Bruntland Commission (1987) *Our Common Future*. Oxford: Oxford University Press.

Caldeira, R. (2008) "My Land, Your Social Transformation': Conflicts Within the Landless People Movement (MST), Rio de Janeiro, Brazil', *Journal of Rural Studies*, 24(2), pp. 150–160.

Carson, R. (1962) *Silent Spring*. Cambridge, MA: Houghton Mifflin.

Carpenter, S.R., Mooney, A., Agard, J., Capistrano, D., DeFries, R., Díaz, S., Dietz, T. et al. (2009) 'Science for managing ecosystem services: Beyond the Millennium Ecosystem Assessment', *Proceedings of the National Academy of Sciences*, 106(5), pp. 1305–1312.

Ceballos, G., Ehrlich, P.R. and Dirzo, R. (2017) 'Biological Annihilation Via the Ongoing Sixth Mass Extinction Signaled by Vertebrate Population Losses and Declines', *Proceedings of the National Academy of Sciences*, 114(30), pp. E6089-E6096.

Cook, R., Cutting, R. and Summers, D (2010) 'If Sustainability Needs New Values, Whose Values? Initial Teacher Training and the Transition to Sustainability', in Jones, P., Selby, D. and Sterling, S. (eds) *Sustainability Education: Perspectives and Practice Across Higher Education*, London: Earthscan pp. 313–327.

Crouch, D. and Ward, K. (1997) *The Allotment: Its Landscape and Culture*. London: Five Leaves.

Daly, H.E. (2005) 'Economics in a Full World', *Scientific American*, 293, pp. 100–109.

DEFRA (Department of Environment, Food and Rural Affairs) (2008) *An introductory guide to valuing ecosystem services*, London: HMSO.

Dempsey, J. and Robertson, M. (2012) 'Ecosystem services: Tensions, impurities, and points of engagement within neoliberalism', *Progress in Human Geography*, 36(6), pp. 758–779.

Dryzek, J. (1997) *The Politics of the Earth*. Oxford: Oxford University Press.

Emery, M. and Flora, C. (2006) 'Spiraling-Up: Mapping community Transformation With Community Capitals Framework', *Community Development*, 37(1), pp. 19–35.

Food for Life Partnership (FfLP) (2011) *Good Food for All: the impact of the Food for Life Partnership*, independent evaluation by NEF, NFER, UWE and Cardiff, Available at: www.foodforlife.org.uk/LinkClick.aspx?fileticket=YyUBCvfUWCc%3d&tabid=310

Gibson-Graham, J.K. (2003) 'Enabling Ethical Economies: Cooperativism and Class', *Critical Sociology*, 29(2), pp. 123–161.

Gibson-Graham, J.K. (2008) 'Diverse Economies: Performative Practices for 'Other Worlds'', *Progress in Human Geography*, 32(5), pp. 613–632.

Haase, D. Larondelle, N. Andersson, E. et al. (2014) 'A quantitative review of urban ecosystem service assessments: concepts, models, and implementation'. *Ambio*, 43(4), pp. 413–433.

Heynen, N. and Robbins, P. (2005) 'The Neoliberalization of Nature: Governance, Privatization, Enclosure and Valuation', *Capitalism Nature Socialism*, 16(1), pp. 5–8.

IAASTD (2008) *Towards Multifunctional Agriculture for Social, Environmental and Economic Sustainability, Issues in Brief. International Assessment of Agricultural Knowledge, Science and Technology for Development*. Available at: www.globalagriculture. org/.../IAASTDBerichte/IssuesBriefMultifunctionality.pdf

Lancaster, O. and Durie, S. (2008) *The Social Return on Investment of Food for Life School Meals in East Ayrshire*. Footprint Consulting: Edinburgh. www.socialvalueuk.org/app/uploads/2016/03/G–Food%20&%20Health-Food%20and%20Health%20Alliance-Website-Resources-FCL%20EAC%20FFL%20SROI%20Technical%20Report%20 12Dec08.pdf

McBurney, S. (1990) *Ecology Into Economics Won't Go: Or Life Is Not a Concept*. Bideford: Green Books.

McClintock, N. (2010) 'Why Farm the City? Theorizing Urban Agriculture Through a Lens of Metabolic Rift', *Cambridge Journal of Regions, Economy and Society*, 3(2), pp. 191–207.

MEA (Millennium Ecosystem Assessment) (2005) *Ecosystems and Human Well-Being: Current State and Trends, Findings of the Condition and Trends Working Group*, Washington DC: Island Press.

Meadows, D.H., Meadows, D.L., Randers, J. and Behrens, W.W. (1976) *The Limits to Growth*. New York: Universe Books.

Michels, A. and De Graaf, L. (2010) 'Examining Citizen Participation: Local Participatory Policy Making and Democracy', *Local Government Studies*, 36(4), pp. 477–491.

Miller, W. (2013) *Allotments and Alternative Food Networks: The Case of Plymouth UK*, Unpublished PhD thesis, University of Plymouth.

MLNR (Ministry of Land and Natural Resources) (1969) *The Thorpe Report: Departmental Committee of Inquiry into Allotments*, Cmnd 4166, London: HMSO.

NEA (2014) UK National Ecosystem Assessment: Follow-On Phase. Available at: www.uknea.unep-wcmc.org/resources

Ostrom, E. (1999) 'Coping with Tragedies of the Commons', *Annual Review of Political Science*, 2(1), pp. 493–535.

Pattie, C., Seyd, P. and Whiteley, P. (2004). *Citizenship in Britain: Values, Participation and Democracy*, Cambridge, MA: Cambridge University Press.

Perrons, D. and Skyers, S. (2003) 'Empowerment Through Participation: Conceptual Explorations and a Case Study' *International Journal of Urban and Regional Research*, 27(2), pp. 265–285.

Pretty, J.N., Ball, A., Lang, T. and Morison, J.I.L. (2005) 'Farm Costs and Food Miles: An Assessment of the Full Cost of the UK Weekly Food Basket', *Food Policy*, 30(1), pp. 1–19.

Rockström, J. et al. (2009) 'A Safe Operating Space for Humanity', *Nature*, 461, p. 472.

Schumacher, E.F. (1973) *Small Is Beautiful: Economics as if People Mattered*, London: Harper & Row.

Scoones, I. (2009) 'Livelihoods Perspectives and Rural Development', *The Journal of Peasant Studies*, 36(1), pp. 171–196.

Seymour, J. (1976) *The Complete Book of Self Sufficiency*. London: Faber &Faber Ltd.

Steel, C. (2008) *Hungry City: How Food Shapes Our Lives*. London: Chatto and Windus.

Tansey, G. and Worsley, T. (1995) *The Food System: A Guide*. London: Routledge.

Taylor, M. (2008) *Living Working Countryside: The Taylor Review of Rural Economy and Affordable Housing*. London: DCLG.

TEEB (The Economics of Ecosystems and Biodiversity) (2018) *Measuring What Matters in Agriculture and Food Systems a Synthesis of the Results and Recommendations of TEEB for Agriculture and Food's Scientific and Economic Foundations Report*. Geneva: UN Environment.

Tudge, C. (2007) *Feeding People Is Easy*. Grosseto: Pari Publishing.

Tudge, C. (2011) *Good Food for Everyone Forever*. Grosseto: Pari Publishing.

Walker, P.A. (2007) 'Political Ecology: Where Is the Politics?' *Progress in Human Geography*, 31(3), pp. 363–369.

Way, T. (2008) *Allotments*. Oxford: Shire Publications.

5 "There's no such thing as a unit of biodiversity"

Contesting value and biodiversity offsetting in England

Guy Crawford

Introduction

In 2011, the UK's Department for Environment, Food and Rural Affairs (DEFRA) released the White Paper titled The Natural Choice: securing the value of nature. The report argues that at present, a plethora of economic and socially beneficial services that derive from the natural environment are not properly accounted for in political and economic decision-making. As part of a drive to rectify this failure, the document details a set of environmental policy proposals, indicative of a broader international movement to green capitalist economies (e.g. UNEP, 2011).

In the paper's foreword, Caroline Spelman MP, who was secretary of state for the environment at the time, states the following:

> This White Paper – the first on the natural environment for over 20 years –
> places the value of nature at the centre of the choices our nation must make:
> to enhance our environment, economic growth and personal wellbeing. By
> properly valuing nature today, we can safeguard the natural areas that we all
> cherish and from which we derive vital services.
>
> (DEFRA, 2011, p. 2)

But how, and through what processes, is it suggested that we "value" nature within our existing political and economic framework? What assumptions are necessary to establish where the "value" of nature is located? And when the signifier of nature is evoked here, what particular vision of this contested concept is being discussed? This chapter explores conflicting discourse surrounding one of DEFRA's proposed policies, biodiversity offsetting, and seeks to highlight how the emergence of new forms of environmental policy involves a semiotic and political struggle between stakeholders to assert a particular vision and understanding of socioecological relations. The discursive focus of this work is not to denounce the materiality of the biophysical realm but rather to explore the contrasting ways that social actors valorise and make sense of relations that comprise socioecological assemblages.

This chapter begins by briefly situating the rise of biodiversity offsetting within the now sizeable literature focused on the relation between neo-liberalism and contemporary forms of environmental policy/governance. The following section details the emergence of biodiversity offsetting in England and provides a brief overview of the proposed policy put forward by DEFRA. Based on a form of qualitative discourse analysis, the bulk of this chapter explores two distinct discourses surrounding DEFRA's biodiversity offsetting policy proposal. Each discourse articulates a particular understanding of socioecological relations that underpin perspectives on how the value of nature should be conceptualised. While the proposed policy is premised on the notion that the coproduction of new sites of biodiversity can negate the loss of sites elsewhere, participants from a government-led public consultation are shown to critique this assumption by interrogating the valuation process that lies at heart of the policy. This critique centres on the importance of acknowledging relations between people and specific places, rather than only attempting to value sites of biodiversity based on an assessment of their habitat type, condition and distinctiveness.

The neo-liberalisation of nature and biodiversity offsetting

In recent years, there has been a proliferation of environmental governance and policy approaches that rely on processes of privatisation, marketisation and commodification to achieve their objectives. Advocates of this policy paradigm argue that such forms can provide a cost-effective and efficient means of addressing the constitutive elements of the contemporary ecological *problematique* that they seek to address. The emergence of this phenomenon has catalysed an ever-expanding collection of scholarly literature that has sought to detail and theorise about how case-specific forms of "neo-liberalised" environmental governance and policy operate (for reviews, see Bakker, 2010, 2015; Castree, 2011).

But how does biodiversity offsetting fit with this phenomenon? Offsetting has been defined as "compensating for losses of biodiversity components at an impact site by generating (or attempting to generate) ecologically equivalent gains, or 'credits', elsewhere" (Maron et al., 2012, p. 141). In the case of the UK, DEFRA's proposed national system for England supports the further expansion of capitalist relations into the biophysical realm through the commodification and marketisation of biodiversity, enabling developers to purchase units that represent expected biodiversity gains. Further detail about the nature of the proposed system is provided in the following section.

The maturing field of "neo-liberal natures" has done much to demystify and critically analyse the processes and outcomes of contemporary forms of environmental policy and governance. Indeed, Carver (this volume) provides a rigorous application of these theoretical frameworks to unpick how exactly the process of bringing nature to market works in the case of UK biodiversity offsetting practices. However, the political-economic work that constitutes much of this field has been criticised for failing to conceptualise contestation surrounding the emergence of "neo-liberal natures" as more than merely political-economic struggles over nature "as resource" (Bakker, 2010, p. 716). This is unsurprising given that traditional political economy approaches tend to neglect cultural elements in their analysis of the social world (Sayer, 2001). A fuller account of the emergence of

"neo-liberal natures" might do well to consider the extent to which manifest articulations of the socioecological realm are contested by stakeholders (see McAfee
and Shapiro, 2010).

In the words of Bakker,

> we might search for a more expansive understanding of neoliberal natures as
> the outcome of cultural, social, and psychological – as well as political, eco-
> nomic, ecological – struggles over our understandings of and relationships
> between socio-natures, both human and non-human.

(2010, p. 728)

It is this interest in the social and cultural dimensions of disputes surrounding the
emergence of neo-liberalised forms of environmental policy that frames this work
on the arrival of biodiversity offsetting in England.

With the exception of Robertson's (e.g. 2000, 2004) pioneering work on wetland mitigation banking in the US, biodiversity offsetting has only recently begun
to receive attention from "critical" quarters of the social sciences and humanities
(Apostolopoulou and Adams, 2015; Benabou, 2014; Carver, this volume; Hannis
and Sullivan, 2012; Lockhart, 2015; Seagle, 2012; Sullivan, 2013a, 2013b; Sullivan and Hannis, 2015; Yusoff, 2011).[1] Robertson's (2012) call for greater focus on
the processes of social construction that are expressed in attempts to ascertain the
"value" of nature provides an opening through which to explore the emergence of
neo-liberal natures as cultural struggles over socioecological sense and meaning.

In this chapter, I will focus on two opposing socioecological discourses and
discuss how they feed into debate surrounding the valuation of biodiversity in the
English land-use planning system. The analysis that informs this work is based
on qualitative discourse analysis of 24 submissions by non-state actors to a public consultation on biodiversity offsetting in England,[2] DEFRA's Green Paper on
the proposed policy and 14 semi-structured interviews held with non-state actor
representatives. Discourse analysis took the form of qualitative coding with the
purpose of understanding how actors construed the socioecological realm in relation to the proposed policy.

The emergence of biodiversity offsetting in England

The arrival of the offsetting paradigm discussed here can be traced back to a report
into the state of the UK's wildlife and ecological network, which was commis-
sioned by the New Labour government in 2009.[3] Lawton and colleagues' (2010)
Making Space for Nature argued that conservation areas in the UK were disinte-
grated and lacked sufficient connectivity. The authors recommended that biodiversity offsetting be trialled in the context of the UK, with the purpose of potentially
codifying a formal scheme that would contribute to the aim of enhancing the UK's
ecological network.

After announcing an intent to explore a "voluntary approach to offsetting" in
the 2011 White Paper, The Environment Bank Ltd, a private biodiversity offset
broker, emerged "to help deliver the government policy" (Environment Bank,

2012). In February 2012, in combination with Mission Markets Inc., The Environment Bank launched the UK's first online trading platform for "conservation credit" (Environment Bank, 2012). Later that year, DEFRA embarked on a biodiversity offsetting pilot study held at six sites across England. The approach taken to offsetting varied between sites and involved local planning authority governance and voluntary participation by private actors. This lasted two years and ran from April 2012 to April 2014 (DEFRA, 2013).

In combination with an effort to gain greater understanding of the practicalities of establishing an offsetting scheme, DEFRA released a Green Paper on the policy in September 2013. This document included 38 questions about the potential form of the policy and marked the launch of a public consultation, which ran for nine weeks and ended on 7 November 2013. The findings of this exercise were not released until February 2016 and make only a passing reference to the discursive dichotomy rehearsed in the course of this discussion (see DEFRA, 2016).

Although the Green Paper included a range of possible policy options, in essence, the proposed system would enable developers to offset "unavoidable" ecological losses at project sites by purchasing an equivalent number of units of biodiversity. These units would be sold by parties that agree to undertake additional biodiversity enhancing land management practices for a stated duration, the objective being to achieve "no net loss" to biodiversity from development. Offsets were discussed as a means of facilitating the final compensatory element of an existing mitigation hierarchy for land-use planning, which is embedded in current UK environmental regulation. DEFRA was also upfront about its ambition to develop a system that is "quicker, cheaper and more certain for developers . . . [while avoiding] additional costs to businesses" (DEFRA, 2013, p. 8).

The pilot scheme's resultant commodification of biodiversity enhancing management practice leads to the production of a form of "socio-natural commodity" (Peluso, 2012, p. 81), where the "Cartesian binary" (Moore, 2011, p. 5) is transgressed. As Cavanagh and Benjaminsen (2014) have discussed in relation to the buying and selling of a tonne of CO_2e within carbon markets, the biodiversity unit is concurrently social and natural, in the sense that the circulation of this form of commodity is premised on the cooperation of a constellation of human and nonhuman actors that work collectively to enable the production, measurement and abstraction of biodiversity into a form that is alienable and ripe for monetary exchange (see Carver, this volume). In the words of one academic interviewee, "[t]here is no such thing as a unit of biodiversity". In the following section, I explore how DEFRA's proposed process of valuation corresponds to non-state actors' perceptions of the value of biophysical space.

The contestable concept of value

Before I discuss how DEFRA proposes to value biodiversity under an offsetting system, I must first outline the government's motivation for considering such an approach. The government's green paper opens by stating that

Our economy cannot afford planning processes that deal with biodiversity expensively and inefficiently or block the housing and infrastructure our economy needs to grow. Our environment cannot afford the wrong type of development which eats away at nature. . . . we should look at new ideas that could help it maintain and improve our ecosystems, air, water and soils as they underpin sustainable economic growth in the long term.

(DEFRA, 2013, p. 1)

As this quote demonstrates, the importance placed on ensuring the maintenance and restoration of ecological systems appears to be closely linked to a structural necessity to ensure the continuation of "sustainable" economic growth. In this sense, biodiversity must be safeguarded (or transposed) as it constitutes the "environmental conditions" of production (O'Connor, 1997, cited in Castree, 2010).

The process by which hitherto unimagined and fantastical forms of ecological commodities are brought into being under neo-liberalised forms of environmental policy involves a series of steps that result in a form of "socially necessary abstraction" (Robertson, 2012, p. 389), where value-bearing delineations are drawn to enable their circulation through market exchange. In the case of the proposed system of biodiversity offsetting in England, DEFRA developed a metric for use in the pilot study areas. This device was designed to permit users to ascertain the "value of habitats" (DEFRA, 2013, p. 10) at different sites and enable ecological loss and gain to be quantified. The Green Paper outlined how the pilot metric had been used and stated that the metric could form the basis for a national system of offsetting.

As noted by DEFRA (2012, p. 2), "[b]iodiversity in its entirety is impossible to measure". Therefore, the proposed policy relies on the use of habitat categories as a proxy. As has been shown by Sullivan (2013b), and illustrated in Table 5.1, the metric quantifies value on the basis of three elements: habitat distinctiveness, habitat quality and area (measured in hectares). The distinctiveness of a hectare of habitat is judged to be one of three possible options, each of which denotes a numerical figure: low (2), medium (4) or high (6).[4] In addition, an assessor must determine the habitat quality of the area in question. Once again, there are three possible options, each of which are represented by a numerical figure: poor (1),

Table 5.1 DEFRA's biodiversity offsetting pilot metric

Value of 1 ha in biodiversity units		Habitat distinctiveness		
		Low (2)	Medium (4)	High (6)
Habitat quality	Good (3)	6	12	18
	Moderate (2)	4	8	12
	Poor (1)	2	4	6

Source: DEFRA, 2013

moderate (2) or good (3).[5] Once both these aspects of the valuation process have been completed, the two figures arrived at are then multiplied together, resulting in a final number that represents the value of the site assessed in biodiversity units. The Green Paper states that the assessment could be undertaken in "as a little as 20 minutes" (DEFRA, 2013, p. 14).

As one can see, the process by which the value of a hectare of land is determined relies exclusively on a simplified form of ecological assessment, neatly described by one interviewee as "a tick-box approach to ecology" (conservation organisation representative). Aside from this reductionist portrayal of (socio)ecological systems, a central premise that lies behind this form of abstraction is that it is unproblematic to treat ecological and social spheres as distinct, where the habitat value of a site, and the implied value to society, can be determined by assessing levels of biological diversity by proxy. In this sense, the *value* is understood as inherently located within material socioecological assemblages. Recent work by Robertson and Wainwright (2013, p. 897) has compared this perspective to Ricardo's understanding of human labour, where value is understood to be "inherent in physical things". Such an approach plainly avoids considering how individuals or local community members interact with a biophysical space, directly or indirectly. As noted by one interviewee,

> "my perspective is that biodiversity offsets are a biodiversity policy, so about preserving our biodiversity resources and their existence into the future, not about whether people can visit and see them . . . the metric being a measure of biodiversity loss and gain, not a measure of social welfare".
>
> (sustainable business group representative)

This discourse of treating biodiversity value as separate and inherent was articulated by DEFRA and three other actors included in the sample analysed. In addition to the notion of value as inherent, the metric also enables numerical comparisons across different habitat categories, habitat *distinctiveness* and area. For example, the value of a hectare of good-quality habitat of low distinctiveness (6) is understood as half the value of a hectare of medium-quality habitat of high distinctiveness (12).

However, the vast majority of participants included in the sample disputed DEFRA's narrow focus on a simplified form of taxonomic assessment as a means to ascertain the value of sites of biodiversity. In contrast to the discourse embodied in the valuation process put forward by DEFRA, most stakeholders rejected the notion that the value of biodiversity could be treated in isolation from the social world. In this case, participants implied that biodiversity located within a physical space needs to be conceptualised as part of a broader socioecological configuration, where the human and the nonhuman constitute an interconnected whole. From this perspective, the method by which DEFRA treats the *social* as conceptually distinct from the *ecological* results in a flawed conception of reality, and, consequently, an inadequate framework for understanding the value of sites of biodiversity:

"environment is not just a haven of wildlife and biodiversity. . . . there is an intrinsic relationship with the human population, . . . an unexplainable, incalculable, importance of the environment for people, whether it's for religious, or community, or just kind of communion, in terms of our well-being. And that, that for us is one of our main concerns for biodiversity offsetting. . . . It's all in the name. It's only about biodiversity".

(social and ecological justice NGO representative)

As this quote illustrates, the way these stakeholders discussed the *value* of nature encompassed more than what a focus on the material, biophysical features of demarcated land can ascertain. Qualitative, nonmaterial dimensions are key to understanding the relations that constitute a particular socioecological configuration and thus how human actors value nature. Respondents that adopted this position were critical of DEFRA's failure to consider the role that potential development sites might already play in the communities that live with, experience and value such places. These non-ecological forms of value were sometimes broadly referred to as the social and cultural value, or local community value, of sites in the landscape in which they are situated. However, respondents also provided examples of specific social and cultural use values, such as amenity, heritage, education, mental health, physical health, aesthetic value, sentimental value and spiritual value. These concerns relating to the policy's failure to engage with such notions of value have also been reported in research by Sullivan and Hannis (2015):

"if you take that particular place away from a community, you're actually damaging the community as much as you are damaging the habitat that's being removed because that community values that place in its particular place in the landscape".

(national park charity representative)

The previous quotation neatly exemplifies an important distinction between de re and de dicto forms of valuation. These concepts have been discussed by the philosopher John O'Neill (2015) in relation to the use of biodiversity offsetting and the idea of natural capital more generally. DEFRA's policy proposal relies on a form of de dicto valuation, where the existence of a *category* of habitat is understood to be that which is valuable. In contrast to this, de re valuation refers to the value of a *particular thing*. In the quotation from the national park charity representative, the interviewee is clearly discussing the importance of a specific place and in doing so demonstrates how areas of biodiversity are often valued in a de re sense. In contrast to the inherent, physical and quantitative notion of biodiversity value articulated by DEFRA and its adherents, the qualitative forms of value discussed can be understood only by acknowledging and attempting to understand the local-level relations between human and nonhuman communities that comprise a specific place.

Locating and located value

DEFRA's valuation metric relies on the assumption that the value of socioecological sites to society derives from the quality and distinctiveness of ranked forms of habitat and enables users to "quickly" and "cost-effectively" calculate ecological loss and gain at participating sites. In the case explored here, the majority of participants did not whole-heartedly reject the notion of buying and selling biodiversity units, but many objected to DEFRA's narrow attempt to locate the habitat value of sites while neglecting already-existing forms of value relations in socioecological entanglements.

It is not that a cultural appreciation of nature is unrecognised by the government; one only has to skim through DEFRA's White Paper to encounter allusions to the value of nature in terms of well-being and unaccounted cultural services (DEFRA, 2011). Yet such concern is devoid when valuation is considered in relation to offsetting. Indeed, the very logic of transposing "equivalent" natures to areas currently safeguarded from the hands of developers runs counter to the evidence base of the DEFRA's White Paper, which notes the "entangled nature of our (and other, nonhuman) attachments to particular places, things and processes" (Yusoff, 2011, p. 4).

The Business and Biodiversity Offset Programme (BBOP), of which DEFRA is an advisory group member, is a growing transnational network of private sector organisations, civil society groups, financial institutions and government agencies that seek to promote biodiversity offsetting as a global environmental conservation strategy (BBOP, 2013). BBOP recognise a need to encompass qualitative and cultural dimensions of socioecological relations into biodiversity valuation (BBOP, 2013). In fact, their conception of "no net loss" encompasses "people's use and cultural values associated with biodiversity" (BBOP, 2013, p. 14). But this raises the question, how are developers supposed to offset sociocultural values ascribed to biodiversity (Benabou, 2014)?

If we cast our eyes further afield, to other forms of state-orchestrated biodiversity offsetting policy, such as in Australia, Brazil, Colombia, Mexico and the US, it seems common practice for biodiversity valuation to negate forms of sociocultural value (Hillman and Instone, 2010; Robertson and Wainwright, 2013; Villarroya et al., 2014). Rather, such systems are defined by attempts, which vary in complexity, to measure, quantify and rationalise nonhuman natures, enabling the calculation of the necessary human and nonhuman labour required for the coproduction of "equivalent" forms elsewhere. In this sense, the international rise of biodiversity offsetting can be characterised as a global phenomenon that entrenches related systems and language of valuation while perpetuating a particular understanding of socioecological relations.

It remains to be seen whether, and in what form, a national system of biodiversity offsetting will emerge from the proposals put forward by DEFRA. As Lockhart (2015) has demonstrated, in the broader context of austerity, the rollout of offsetting has been plagued by tensions between establishing a simple, business-friendly system and one accepted as robust in both a regulatory and ecological sense. In the years since the DEFRA pilot ended, participating local authorities have continued to independently experiment with offsets. Furthermore, in January 2018, Natural England (a public government advisory body) conducted a further survey on the original DEFRA valuation metric. These activities suggest that incorporating a national system of offsets into the English land-use planning system remains a future possibility.

However, if such a system were to emerge, it remains unknown whether any attempt will be made to move beyond its current focus to include sociocultural dimensions in biodiversity valuation. It is perhaps telling that Natural England's recent survey made no reference to these discussions. Indeed, given the government's stated preference to develop a system that reduces complexity and costs for developers, it seems doubtful that an expanded and more inclusive approach to determining value would be welcomed.

Conclusion

The UK government's experimentation with offsetting as a means of ameliorating the metabolic relationship between biodiversity and growth in the development sector constitutes a new form of "nature as accumulation strategy" (Smith, 2007), where a hitherto untapped aspect of the biophysical realm assumes a commodity form. Confronted with the repercussions of the North Atlantic financial crisis of 2007–2008 and an ongoing environmental crisis in the form of continued biodiversity loss, the formal state, in its role as the prime regulator of capital accumulation (O'Connor, 1997, cited in Castree, 2010), sought to act. The proposed policy would address two objectives:

1 The reduction of regulatory burdens on the development sector in the hope of catalysing increased economic activity
2 The assignment of monetary value to "additional" biodiversity enhancing land management practices in an attempt to maintain and enhance stocks of "natural capital" across England

This chapter has sought to highlight the extent to which DEFRA's proposed system of biodiversity valuation, and the underlying assumptions relating to the implied value of nature to society, was greatly disputed by the majority of non-state actors included in this research. In contrast to the notion of locating value within material socioecological configurations, which help to maintain economic activity, non-state actors asserted the importance of understanding the value of biophysical space as being derived from "place-based cultural values of specific localities".[6]

In *The Natural Choice: securing the value of nature* (2011), the DEFRA defines biodiversity as follows:

> Biodiversity is life. We are part of it and we depend on it for our food, livelihoods and wellbeing. It is the term used to describe the variety of all life on Earth.
>
> (DEFRA, 2011, p. 17)

Following DEFRA, if we understand the human as a constituent component of biodiversity, surely it follows that biodiversity valuation should include formal considerations of the relations, values and significance given to the places that comprise the landscapes in which human and nonhuman natures relate and reside.

Notes

1 Sullivan and Hannis's (2015) recent study also explores the discursive nature of debate surrounding biodiversity offsetting in England. While this book chapter solely analyses contrasting positions on the valuation of biodiversity, Sullivan and Hannis (2015) provide a broader analysis of value frames that underpin perspectives on many facets of the policy proposal.
2 Non-state actors included civil society organisations (15), private sector actors (8) and a joint submission by two academics.
3 A form of offsetting exists under the EU Habitats Directive (see Sullivan, 2013b for a case study).
4   Distinctiveness reflects, among other factors, the rarity of the habitat concerned (at local, regional, national and international scales) and the degree to which it supports species rarely found in other habitats. Guidance has been provided alongside the pilot, setting out the distinctiveness rating for different habitat types.

(DEFRA, 2013, 10)

5 This assessment is based on a guidance manual produced by Natural England titled Higher Level Stewardship: Farm Environment Plan (FEP) Manual (2010) (DEFRA, 2013).
6 This quote was taken from a joint consultation submission by two academics.

References

Apostolopoulou, E. and Adams, W.M. (2015) 'Neoliberal Capitalism and Conservation in the Post-crisis Era: The Dialectics of "Green" and "Un-green" Grabbing in Greece and the UK', *Antipode*, 47(1), pp. 15–35.
Bakker, K. (2010) 'The Limits of 'Neoliberal Natures': Debating Green Neoliberalism', *Progress in Human Geography*, 34(6), pp. 715–735.
Bakker, K. (2015) 'Neoliberalization of Nature', in Perreault, T., Bridge, G. and McCarthy, J. (eds) *The Routledge Handbook of Political Ecology*. Oxford: Routledge, pp. 446–456.
BBOP. (2013) *To No Net Loss and Beyond: An Overview of the Business and Biodiversity Offsets Programme (BBOP)*. Washington, DC. Available at: www.forest-trends.org/documents/files/doc_3319.pdf.
Benabou, S. (2014) 'Making Up for Lost Nature? A Critical Review of the International Development of Voluntary Biodiversity Offsets', *Environment and Society: Advances in Research*, 5(1), pp. 103–123.
Castree, N. (2010) 'Neoliberalism and the Biophysical Environment 2: Theorising the Neoliberalisation of Nature', *Geography Compass*, 4(12), pp. 1734–1746.
Castree, N. (2011) 'Neoliberalism and the Biophysical Environment 3: Putting Theory into Practice', *Geography Compass*, 5(1), pp. 35–49.
Cavanagh, C. and Benjaminsen, T.A. (2014) 'Virtual Nature, Violent Accumulation: The 'Spectacular Failure' of Offsetting at a Ugandan National Park', *Geoforum*, 56, pp. 55–65.

DEFRA. (2011) *The Natural Choice: Securing the Value of Nature*. London: Department for Environment, Food, and Rural Affairs. Available at: www.gov.uk/government/uploads/system/uploads/attachment_data/file/228842/8082.pdf.

DEFRA. (2012) *Biodiversity Offsetting Pilots Technical Paper: The Metric for the Biodiversity Offsetting Pilot in England*. London: Department for Environment, Agriculture and Rural Affairs. Available at: www.gov.uk/government/uploads/system/uploads/attachment_data/file/69531/pb13745-bio-technical-paper.pdf.

DEFRA. (2013) *Biodiversity Offsetting Pilots in England Green Paper*. London: Department for Environment, Food and Rural Affairs. Available at: https://consult.defra.gov.uk/biodiversity/biodiversity_offsetting.

DEFRA. (2016) *Consultation on Biodiversity Offsetting in England: Summary of Responses*. London: Department for Environment, Agriculture and Rural Affairs. Available at: https://assets.publishing.service.gov.uk/government/uploads/system/uploads/attachment_data/file/501240/biodiversity-offsetting-consult-sum-resp.pdf.

Environment Bank. (2012). *Environment Bank and Mission Markets Launch Online Conservation Credit Platform*. Available at: www.environmentbank.com/docs/Environment%20Bank-Mission%20markets%20release%20feb2012.pdf.

Hannis, M. and Sullivan, S. (2012) *Offsetting Nature? Habitat Banking and Biodiversity Offsets in the English Land Use Planning System*. Weymouth: Green House.

Hillman, M. and Instone, L. (2010) 'Legislating Nature for Biodiversity Offsets in New South Wales, Australia', *Social & Cultural Geography*, 11(5), pp. 411–431.

Lawton, J.H., Brotherton, P.N.M., Brown, V.K., Elphick, C., Fitter, A.H., Forshaw, J. et al. (2010) *Making Space for Nature: A Review of England's Wildlife Sites and Ecological Network*. Report to DEFRA. Available at: http://archive.defra.gov.uk/environment/biodiversity/documents/201009space-for-nature.pdf.

Lockhart, A. (2015) 'Developing an Offsetting Programme: Tensions, Dilemmas and Difficulties in Biodiversity Market-Making in England', *Environmental Conservation*, 42(4), pp. 335–344.

Maron, M., Hobb, R.J., Moilanen, A., Matthews, J.W., Christie, K., Gardner, T.A., Keith, D.A., Lidenmayer, D.B. and McAlpine, C.A. (2012) 'Faustian Bargains? Restoration Realities in the Context of Biodiversity Offset Policies', *Biological Conservation*, 155, pp. 141–148.

McAfee, K. and Shapiro, E. (2010) 'Payments for Ecosystem Services in Mexico: Nature, Neoliberalism, Social Movements, and the State', *Annals of the Association of American Geographers*, 100(3), pp. 579–599.

Moore, J.W. (2011) 'Transcending the Metabolic Rift: A Theory of Crises in the Capitalist World-Ecology', *The Journal of Peasant Studies*, 38(1), pp. 1–46.

O'Connor, J. (1997) *Natural Causes: Essays in Ecological Marxism*. London: The Guilford Press.

O'Neill, J. (2015) 'Sustainability', in Moellendorf, D. and Widdows, H. (eds) *The Routledge Handbook of Global Ethics*. London: Routledge, pp. 401–415.

Peluso, N.L. (2012) 'What's Nature Got to Do With It? A Situated Historical Perspective on Socio-natural Commodities', *Development and Change*, 43(1), pp. 79–104.

Robertson M (2000) 'No Net Loss: Wetland Restoration and the Incomplete Capitalisation of Nature', *Antipode*, 32(4), pp. 463–493.

Robertson, M. (2004) 'The Neoliberalization of Ecosystem Services: Wetland Mitigation Banking and Problems in Environmental Governance', *Geoforum*, 35(3), pp. 361–373.

Robertson, M. (2012) 'Measurement and Alienation: Making a World of Ecosystem Services', *Transactions of the Institute of British Geographers*, 37(3), pp. 386–401.

Robertson, M. and Wainwright, J.D. (2013) 'The Value of Nature to the State', *Annals of the Association of American Geographers*, 103(4), pp. 890–905.

Sayer, A. (2001) 'For a Critical Cultural Political Economy', *Antipode*, 33(4), pp. 687–708.

Seagle, C. (2012) 'Inverting the Impacts: Mining, Conservation and Sustainability Claims Near the Rio Tinto/QMM Ilmenite Mine in Southeast Madagascar', *The Journal of Peasant Studies*, 39(2), pp. 447–477.

Smith, N. (2007) 'Nature as Accumulation Strategy', in Leys, C and Panitch, L (eds) *The Socialist Register 2007: Coming to Terms With Nature*. London: Merlin, pp. 1–21.

Sullivan, S. (2013a) 'Banking Nature? The Spectacular Financialisation of Environmental Conservation', *Antipode*, 45(1), pp. 198–217.

Sullivan, S. (2013b) 'After the Green Rush?: Biodiversity Offsets, Uranium Power and the 'Calculus of Casualties' in Greening Growth', *Human Geography*, 6(1), pp. 80–101.

Sullivan, S. and Hannis, M. (2015) 'Nets and Frames, Losses and Gains: Value Struggles in Engagements With Biodiversity Offsetting Policy in England', *Ecosystem Services*, 15, pp. 162–173.

UNEP. (2011) *Towards a Green Economy: Pathways to Sustainable Development and Poverty Eradication*. United Nations Environment Programme. Available at: www.unep. org/greeneconomy/Portals/88/documents/ger/ger_final_dec_2011/Green%20Economy Report_Final_Dec2011.pdf.

Villarroya, A., Barros, A.C. and Kiesecker, J. (2014) 'Policy Development for Environmental Licensing and Biodiversity Offsets in Latin America', *PLSO One*, 9(9), p. e107144.

Yusoff, K. (2011) 'The Valuation of Nature: The Natural Choice White Paper', *Radical Philosophy*, 170, pp. 2–7.

6 Commensuration as value making

Transforming nature in English biodiversity offsetting under the DEFRA metric

Louise Carver

Introduction

While efforts to ascertain values for nonhuman nature and natural environments are nothing new in theoretical and increasingly policy fields, nascent scholarly interest in the modes and methods of such practices are coalescing at the act of valuation itself (Helgesson and Muniesa, 2013). Within this, there is scope to illuminate the normative assumptions and expert practices that become "black boxed in diverse technical infrastructures" (Kjellberg et al., 2013, p. 17). For example, valuing nature with biodiversity offsetting is frequently critiqued as the commodification of biotic entities through the entangled logics of capitalism and conservation (Dauguet, 2015; Sullivan, 2010, 2012). But what does it actually mean to attribute exchange value to nature by using biodiversity offsetting even if, as some have pointed out, the mechanism rarely produces a true "market" (Boisvert, 2015) or where its "goods" might never actually be sold? To clarify what it means to create new economic value from the ecological functions of bio-diversity, I query, if there is value in standing stocks of conserved or re-created biodiversity, then in what ways is this being realised? Indeed, where, when and how is it located?

This chapter engages with these questions through a detailed investigation of the UK Department for Environment, Food and Rural Affairs (DEFRA) biodiversity metric, as a "calculative device" (Callon and Muniesa, 2005; Callon, 2007). Set against a relative paucity of case material in a fast-developing European biodiversity offsetting policy arena (but see Dauguet, 2015), the DEFRA metric enacts new value through its function to make commensurate the biodiversity values between sites of ecological loss and gain in development and planning in England. It illuminates the iterative layers of value production achieved through the metric's sequential transformation of nature via Excel spreadsheets, so that it is newly realised as equivalent to other kinds of natures and therefore subsequently liquid exchange value. The chapter draws from the performative tradition in economic sociology (Callon, 2007; Calışkan and Callon, 2009, 2010; Mackensie and Millo, 2003; MacKensie, 2006), which begins, not with the assumption of an external a priori existence of the "economic X" but instead asks how it is that the economic is made and performed under processes of economisation and marketisation (Çalışkan and Callon, 2010). It argues that under English biodiversity offsetting policy, the DEFRA metric interacts with the physical landscape subject to anthropogenic changes, to reorganise its ecological content as valuable in the eyes of planning authorities and other interested parties. In the spirit of this edition's collective endeavour to locate value, spatially or otherwise, and discern its properties, I propose that a primary way biodiversity value is realised is through colliding with sites allocated for modification with development and the concurrent conservation work provided for biodiversity offsetting. Ecological assemblages are individuated and abstracted from material landscapes, and valorised thus, because of their coincidence with development and planning spatialities and temporalities. Put simply, biodiversity value denoted in novel units and credits exists, in development and at offset sites, because it is measured (Mitchell, 2008). The metric first locates biodiversity value through the spatial and temporal geographies of construction and development through the metric and subsequently circulates it as exchange value on the catalogues and digital registers of institutional actors engaged in biodiversity offsetting.

The empirical cases drawn on here are based on 18 months of fieldwork with two DEFRA biodiversity offsetting local planning authority pilot sites. This research involved following specific offset cases through the English planning system using repeat visits and interviews with a sample of eight to ten types of stakeholders involved in each case study. A detailed analysis of a range of primary and secondary textual documents from the public and private domains gives additional support to the interview data. The chapter draws from Castree's (2003) Marxian typology of commodification processes and illustrates biodiversity offsetting as a commodification practice through exploring some empirical examples of how the metric functions to render distinct ecologies commensurate with and equivalent to one another. It proceeds by laying out the processes of quantification, abstraction and individuation in revealing how it is that the ecological is translated into the economic.

Locating new value in nature; or the act of "making equivalent"

I borrow the expression "calculative devices" (Callon and Muniesa, 2005; Callon, 2007) as a framework to investigate an emerging suite of quantification technologies that are used to influence social outcomes by public and private actors. Calculative devices, such as equations, formulas or scorecards, or in this case the DEFRA biodiversity metric, both shape and determine the valuation process of any particular calculative technology – for example, biodiversity offsetting (Bracking et al., 2014). The power of the device is that it produces new, calculated entities, which can be counted and accounted for, costed and circulated as new commodities. This section details the iterative layers of value creation underlying the transformative processes in which biodiversity value is first constructed as a new conceptual category, stabilised as a commodity and thereby "made" into a unit of exchange. In other words, it considers the processes through which offsetting enacts the commodification of biodiversity function as a way of producing and then realising this new value in an economic sphere.

The production of new value within the DEFRA metric derives principally from two general processes that perform notional commensurability between biotic entities. The first of these processes recodifies ecological information into numerical scores, and the second assigns these scores a pecuniary value (Sullivan, 2010). Here I follow Castree's (2003) typological review of commodification processes and draw from Thomas, who proposes that "the commodity status of a thing is not intrinsic, it is assigned" (1992, p. 28), underlining the thesis that value is performed, not discovered (Mackensie, 2006). Castree suggests that "the question then is not what is a commodity but what kind of characteristics do things take on when they become commodities?" (2003, p. 277). Following Castree, we can see how it is that biodiversity as a general abstracted category (and its constitutive animate components) is symbolically transformed to operate as a commodity for exchange.

What, then, qualifies biodiversity offsetting specifically as a form of commodification? Castree argues that at the most abstract level, capitalist commodification is

> a process that ensures qualitatively distinct things are rendered equivalent and sellable through the medium of money. Particular commodity-bodies (use values) are thus commensurated and take on the general quality of exchange value.
>
> (2003, p. 278)

Thus, it is the particular acts that render temporally and geographically specific natures commensurable with other natures and ultimately with pecuniary value, which underpins the commodification process. As I will go on to show, the financial valuation stage is not in fact even required to classify the process as one of commodification. Assigning numerical signifiers as unit values constitutes the first step, which is followed by the construction of notional equivalence in performing exchange value. In this, however, processes of abstraction and individuation occur as constitutive to these *two* primary stages of commensuration: quantification and monetisation. Before I turn to abstraction and individuation, I must first examine these two *general* performances: constructing numerical score cards and designing pricing methods around those numbers.

Creating and pricing numerical continuums

The first stage of the process in constructing a single unit of biodiversity value constitutes the translation of nature, and knowledge about nature, into a medium that makes it commensurable and comparable along a uniform continuum of equivalent values. Crawford (this volume) introduces us to the basic design guidelines of the DEFRA metric, which multiplies scores across habitat distinctiveness, condition and size to produce a numerical biodiversity value. In practice, this begins with a private consultant ecologist in service to a developer client, completing a phase 1 habitat survey, as would happen during the creation of any other ecological assessment report to be submitted in support of a planning application. As a desk-based exercise, before sending to the local planning authority, the ecologist translates this information into the biodiversity impact assessment (BIA) Excel spreadsheet using the DEFRA metric formulas that categorise spatial areas of the site, on separate rows of the sheet, into habitat types, each with a corresponding code, description, size in hectares and category scores for distinctiveness and condition. The numbers in each row are subsequently multiplied to produce the habitat biodiversity value for each coded spatial arrangement of on-site habitat categories subject to development. It is the nature of the Excel spreadsheet here that I wish to emphasise in its function as a calculative device to SUM rows of geographically bounded, labelled and rated habitat types as measures of quality and rarity. This signified "biodiversity value" travels along three stages of an Excel calculation, before being materialised as financial compensation payments from developer to the local authority or broker, such as the Environment Bank Ltd, and administered in down payments to an offset provider over a 30 year period. The DEFRA biodiversity metric, as practised through Excel spreadsheet formulas facilitates the articulation of scientific information into legal and economic fields (cf. Robertson, 2006) through the production of numerical signifiers as proxies for qualitative, ecological assemblages.

The function of creating a numerical continuum is illustrated by one particular planning application for development that falls under the DEFRA pilot study. A proposed development of a plot of over 14.8 hectares would see it transformed from largely agricultural use for livestock grazing to 236 residential properties, a football and bowling club and commercial buildings. Almost half of the on-site mitigation work (i.e. habitat creation to cancel out the ecological debt imposed through development) was to be delivered by creating football pitches. Lest the reader be wondering what strategic value football pitches serve to the recovery of biodiversity in England, their presence in this development contributed a high aggregated numerical score of biodiversity *unit* value, and indeed the largest single source of habitat value for the development, by virtue of their size alone. Scoring low for both habitat distinctiveness (2) and habitat condition (1), it is the scale of eight football pitches that make up more than 40% of total on-site mitigation and habitat creation, thus reducing the perceived total biodiversity impact of that development. This reduction of qualitative and quantitative values that are already various simplified abstractions (habitat condition, distinctiveness and spatial size) to one uniform numerical unit makes the idea of net gains and losses coherent. Indeed, the conjuring of a comparative relational quality between highly heterogeneous entities rendering nature measureable and thus rationally manageable permeates the appeal of biodiversity offsetting in policy circles. As Julia Marton-Lefèvre, former director general of the international conservation agency IUCN expressed at the World Forum for Natural Capital in 2013, "business and conservation have been two different camps counting different things, and now for the first time they are able to speak the same language".[1]

The second layer of commensuration, through a function of price, embodies the moment that the commodification process is materialised even if the prior stages demonstrate functional commodity production in the absence of money. The financial costs of the offset are derived from the overall cost of providing habitat to the value of the units required and are to be set by the offset provider (the biodiversity unit vendor). This is therefore highly variable and dependent on wider market conditions such as land prices. The financial costs of the offset are made up from "the average between a calculated cost of purchase or rental of a 1-hectare piece of land, certain management and some capital works to that 1 hectare" (Interview, local planning authority pilot lead, November 2013).

With monetary valuation underwriting the new liquidity of biodiversity through symbolic commensuration and physical exchange, it is the transmutation of ecological qualities into a format in which they are susceptible to quantitative financial valuation that completes the second principal layer of exchange value creation. Of course, this is possible only due to series of discursive, semiotic and subsequently material reconfigurations to ecological entities and places. Subprocesses of abstraction and individuation, as I illustrate in the following sections, render such entities and places alienable and thus amenable to being commensurate and therefore valuable. Individuation and abstraction are shown to be constitutive of, and implicit in, the two higher stages of commensurability involved in the production of value in biodiversity offsetting.

Individuation

Individuation refers to the "representational and physical act of separating a specific thing or entity from its supporting context" (Castree, 2003, p. 280). It follows that the "material boundaries are demarcated" such that the thing can be bought, sold and exchanged, by "equally bounded groups or individuals" (ibid., p. 280). Therefore, "it involves a discursive and practical 'cut' into the seamless complexity of the world in order to name discrete 'noun-chunks' of reality are deemed to be socially useful" (ibid., p. 280). In biodiversity offsetting, these are units that conceptually alter ecologically expansive spaces into "market-friendly ones" (Robertson, 2006).

In considering the function of individuation in the creation of biodiversity value in biodiversity offsetting, an example may be drawn from the "discursive and practical cut" used to produce a single distinct hectare of cirl bunting (*Emberiza cirlus*) breeding territory. Cirl buntings were once a common farmland bird almost lost to the UK in the 20th century. It is now almost entirely restricted to parts of southern England, due largely to habitat loss and agricultural intensification. To compensate for the loss of their habitat through development, using the offset mechanism, planners must first isolate one hectare of breeding territory so that the abstracted value of the loss can be quantified, monitored and rendered equivalent to other 1 hectare units of compensation to be provided. A local planning officer notes that

"There are question marks as to what the trigger is for that 1 hectare. A breeding territory is a 250 m radius from a singing male, who will generally be singing fairly near or within a 250 m radius of his nest. The problem is that when you record him, he may be on the 250 m mark away from his nest. We then put a theoretical 250 m buffer zone around this singing male and say that *that* is his breeding territory. A 250 m radius creates, I think it's about a 19 or 21 hectare area, so if you lose 5% of that area, that equates to about 1 hectare."

(Interview, local planning authority pilot lead, November 2013)

The 16th-century Enclosure Act saw the transformation of commons landscapes that were sliced, diced and delimited according to the property rights and economic productivity of land for its owners. Similarly, today, landowners of plots allocated as an offset receptor site or habitat bank are expected to divide this up along the new economic and legal boundaries of individual biodiversity offsets. The administrative boundaries of these plots define newly individuated productive landscapes that generate value by providing a specific number of conservation credits of delineated ecosystem services through biodiversity credits. These credits are subsequently exchanged with the similarly individuated biodiversity debits carried by a range of geographically divergent developments and are called buyers.

Abstraction

Castree (2003) distinguishes between two kinds of abstraction: functional and spatial. Functional abstraction is understood to be "where the qualitative specificity of any individualised thing is assimilated to the qualitative homogeneity of a broader type or process" (Castree, 2003, p. 281). It follows that the aggregation of subcomponents during the process of standardisation inherent in the biodiversity metric and the production of biodiversity units through abstraction with a numerical proxy are ones of potentially extreme simplification in the context of pronounced complexity and indeterminate uncertainty. Consequently, Sullivan asks whether these calculations and simplifications are subsequently "profoundly un-ecological" (2012, p. 31).

For example, a development that was approved on the edge of SSSI woodland uses the locally adapted metric formulas to establish what value, in unit terms, constitutes the site baseline and will subsequently be lost through development and necessarily compensated for. The site in question comprises 0.8 ha of lowland heathland, which is a priority habitat of high distinctiveness (score: 6) but in poor condition (score: 1), meaning 4.8 biodiversity units will be lost, expressed in the following equation:

$$0.8 \times 6 \times 1 = 4.8$$

The offset habitat to be restored as compensation for the impacted site has a baseline value of 1.2 units, which is calculated by 0.2 ha, with a distinctiveness score of 6 and a low condition score of 1.

$$0.2 \times 6 \times 1 = 1.2$$

It is proposed that the value of the offset site can be doubled to 2.4 units by improving the condition of the lowland heathland from 1 to 2 units, conveyed qualitatively by improving its condition from poor to moderate. Two questions arise in respect to this small illustrative example. The first relates to the distinctive and material differences between a poor condition a and moderate condition of lowland heathland such that the habitat numerical value can be doubled by a category upgrade. The condition assessment criteria are devised at the local planning authority level in accordance with Natural England and are derived from the Farm Environment Plan (FEP) Manual (Natural England, 2010). Condition assessments for lowland heathland habitats are by and large characterised by the coverage and frequency of shrubs and heather above a certain percentage threshold and undesirable shrubs and trees below a percentage threshold, but not graded bands according to specific percentages of coverage (Natural England, 2010). The qualitative and quantitative precision in establishing the condition of a lowland heath according to three graduated bands of good, moderate and poor along uniform axes of heather coverage and shrub maturity shape the category outcomes in ways that send signals back along the ecological and financial value judgements of the calculation. Yet they seem to derive from a range of assessments that are directed by not insignificantly generalised and idiosyncratic guidelines and that disguise incertitude, subjective albeit expert judgements and complexities in seemingly stable numerical values.

The second question relates to moving around numerical units of chopped-up habitat to produce a quantitative "net" gain in overall biophysical indicators. For example, the heathland of this offset example is also ecologically proximate to and part of a wider nightjar nesting habitat. Where the local planning officer overseeing this case suggested the heathland habitat offset will be relatively simple to obtain, they reported that finding nightjar compensation was "proving less easy" (local planning authority pilot lead, July 2014). The officer hopes "both can be accommodated by the same offset", illustrating that by design, they need not necessarily be so. Thus, the functional abstraction produces value specifically from the "conceptual disaggregation of species from ecosystem fabrics and their re-embedding within calculative rationalities of quantification" (Sullivan, 2012, p. 25).

Standardising ecological qualities is necessary to make things comparable but of course comes with difficulties in so far as it hides categorically irrelevant divergences and complexities (Lamont, 2012). This is a conflict one might reasonably expect in an area as complex, relational and indeterminate as ecosystems and wildlife habitats, where consensus on discrete, stable and coherent categories and the business of isolating who is doing what to whom (Yusoff, 2011) may prove elusive. Standardisation is intrinsic to the preparation of a market so as to achieve a feasible degree of substitutability between commodities, thus facilitating a trade (Callon and Muniesa, 2005). This proves to be the case for biodiversity exchange processes, as in any other commodity market. By placing goods in a frame with other goods and establishing a relational connection between them, new types of classifications and calculations can occur (Callon and Muniesa, 2005), and value is thereby produced.

The second form of abstraction in the process of commensuration between different natures is spatial, for which functional abstraction is a precondition (Castree, 2003). Castree writes that spatial abstraction involves "any individualized thing in one place being treated as *really the same* as an apparently similar thing located elsewhere" (2003, p. 281). The construction of functionally abstract units of biodiversity means that the spatial specificity of one habitat or place can be substituted for another. An ecological consultant concerned with the implications of spatial abstraction articulates concerns over this scientific approach:

> "Between the site of ecological loss and gain, the geographical, edaphic and microclimatic factors alone of two separated sites mean the ecology would not be the same. And to suggest (as habitat banking implies) that you can offset a site in one place with another somewhere else (often a long way away) is just a demonstration that the basic science has not been understood – and the obfuscating metrics produced by offset proponents just serve to add a false patina to make it look like science".
>
> (private consultant ecologist, February 2014)

Discussion

This chapter has presented some brief empirical renderings of the complexities in trying to collapse the demarcations between the ecological and the economic. In so doing, the former is rendered liquid and fluid as newly materialised as exchange value. This critical capitalist shift in the logics of conservation has been articulated elsewhere as "nature on the move": a nature that is no longer static but mobile, reflecting the classical Marxian concept of "value as process" (Büscher, 2013, p. 21, citing Marx, 1976, p. 256). Exchange value is created with each layer of imposed commensuration in the steps of an offset calculation and constitutes an epistemic and ontological transformation to the primary nature that is affected, constructing credit and debt value that circulates on the registers and inventories of institutional offset practitioners and local authorities. By rendering delineated units of biodiversity disconnected from specific places as mobile values in (pseudo-)market space, biodiversity is at last compatible in form and function with the physical and economic geographies of development.

As well as "locating" the nascent value produced through the metric as a calculative entity, this chapter questions the transformative potential of the new "valuation" paradigm in conservation and planning governance. For example, in considering the imaginative possibilities of valuing nature with unit scores and economic proxies, it asks whether the processes associated with rendering biodiversity exchangeable and commensurable may obfuscate more meaningful conceptualisations of these biotic entities, relationships and places. Might their disclosure by economic valuation leave them more marginalised in social decision-making than they were before (Yusoff, 2011, p. 4)? How might policymakers and conservation actors ameliorate the difficulties of articulating appropriate ways to value flourishing biodiversity and ecological systems as both integral and instrumental to human well-being *as well as* intrinsic, specific ends in their own right (Latour, 2005)? In the spirit of pragmatism, conservation biologists and policymakers are understandably searching for solutions that are compatible with preponderant economic and political priorities (Sandbrook et al., 2013) and oriented towards rationalistic cost-benefit analyses along an economic continuum. But the ideological, conceptual and imaginative work that these shifts are performing and to whom in particular this will be "of value" deserve further attention.

Hopefully, when considering the methodological challenges of valuation, one remains attentive to the broader structuring logics of the political realities of the valuation context, lest such efforts fall foul of fulfilling the "tragedy of the *well-intentioned* valuation" through the compromises that these require (Gómez-Baggethun and Ruiz-Pérez, 2011, p. 624). In calling for greater awareness of the sociopolitical context of valuation and its desired political goals, Gómez-Baggethun and Ruiz-Pérez (2011) elucidate the pitfalls of well-meaning, albeit politically naïve, approaches to the valuation of nature that may all too frequently confirm and reinforce the conditions of ecological decline through acting in primary service to the higher neo-liberal political orders of worth for economic growth and efficiencies. Both the DEFRA Natural Environment White Paper and Natural England illuminate the political conditions for environmental governance in England at the present time as one in service to, and providing support for, *sustainable development* and the growth of the UK economy. This guidance is reflected in the everyday working culture of Natural England, as a senior ecologist articulates: "at the moment, there is a big emphasis on growth, and as an organization, we have a responsibility to look for solutions ensuring the environment is not in conflict with that" (Natural England, November 2014).

Conclusion

Biodiversity offsetting policy in England and its valuation practices do not derive from a linear institutional process but instead are being realised, tested, contested and iterated via a range of "complex cognitive, analytical, discursive, political, institutional and material devices" that are acting together to reshape conservation norms and practices (Kallis et al., 2013, p. 99). Where the previous chapter introduced the reader to the epistemological origins of the contested framings and policy narratives, this chapter presents the DEFRA biodiversity offsetting metric as a calculative technology that performs valued ecological entities in the nascent biodiversity markets in England. These calculated entities perform and conform with value frames that act primarily in service to broader political-economic priorities for economic growth through the built environment, by detaching these values from specific places and instead attaching them to development spatialities. The entities do not easily correspond to the actual material reality of natures in situ or represent an accurate commensurability with these, nor could they. The question remains whether aspirations for valuing nature through biodiversity offsetting is capable of fulfilling the expectations of those involved or at least the avowed good intent of the proponents of the approach in order to deliver a renewed and

88 *Louise Carver*

invigorated array of socioenvironmental relations. After all, the valuing-nature discourse, experienced by some as an optimistic and transformative opportunity, is avowedly in service to this greater effort. Or perhaps the symbolic violence of the metric, through producing abstracted and individualised credit and debit biodiversity units on digital inventories, renders nature even more disposable and alienable than before such "valuation".

Note

1 Opening address at the World Forum for Natural Capital, Edinburgh, November 2013.

References

Boisvert, V. (2015) 'Conservation banking mechanisms and the economization of nature: An institutional analysis', *Ecosystem Services*, 15, pp. 134–142.

Bracking, S. Bond, P. and Brockington, D., et al. (2014) *Initial Research Design: Human, Non-Human and Environmental Value Systems: An Impossible Frontier?* Leverhulme Centre for the Study of Value Working Paper Series No. 1, Manchester.

Büscher, B. (2013) 'Nature on the Move I: The Value and Circulation of Liquid Nature and the Emergence of Fictitious Conservation', *New Proposals: Journal of Marxism and Interdisciplinary Inquiry*, 6(1–2), pp. 20–36.

Çalışkan, K. and Callon, M. (2009) 'Economization, Part 1: Shifting Attention From the Economy Towards Processes of Economization', *Economy and Society*, 38(3), pp. 369–398.

Çalışkan, K. and Callon, M. (2010) 'Economization, Part 2: A Research Programme for the Study of Markets', *Economy and Society*, 39(1), pp. 1–32.

Callon, M. (2007) 'What Does it mean to Say Economics is Performative?' in MacKensie, D., Muniesa, F. and Siu, L. (eds) *Do Economists Make Markets? On the Performativity of Economics*. Princeton, NJ: Princeton University Press.

Callon, M. and Muniesa, F. (2005) 'Economic Markets as Calculative Collective Devices', *Organisation Studies*, 26(8), pp. 1229–1250.

Castree, N. (2003) 'Commodifying What Nature?' *Progress in Human Geography*, 27(3), pp. 273–297.

Dauguet, B. (2015) 'Biodiversity Offsetting as a Commodification Process: A French Case Study as a Concrete Example', *Biological Conservation*, 192, pp. 533–540.

Gómez-Baggethun, E. and Ruiz Pérez, M. (2011) 'Economic Valuation and the Commodification of Ecosystem Services', *Progress in Physical Geography*, 35(5), pp. 613–628.

Kallis, G., Gómez-Baggethun, E. and Zografos, C. (2013) 'To value or Not to Value? That Is Not the Question', *Ecological Economics*, 94, pp. 97–105.

Kjellberg, H. Mallard, A. and Arjaliès, D.-L., et al. (2013) 'Valuation studies? Our collective two cents', *Valuation Studies*, 1(1), pp. 11–30.

Lamont, M. (2012) 'Toward a Comparative Sociology of Value and Evaluation', *Annual Review of Sociology*, 38(1), pp. 201–221.

Latour, B. (2005) *Reassembling the social: An introduction to Actor-Network Theory*. Oxford: Oxford University Press.

Mackensie, D. (2006) *An Engine, Not a Camera: How Financial Models Shape Markets*. Cambridge, MA: MIT Press.

Mackensie, D. and Millo, Y. (2003) 'Constructing a Market, Performing a Theory: The Historical Sociology of a Financial Derivatives Exchange', *American Journal of Sociology*, 109(1), pp. 07–145.

Marx, K. (1976) *Capital. Volume I*. London: Penguin Books.

Mitchell, T. (2008) 'Rethinking Economy', *Geoforum*, 39(3), pp. 1116–1121.

Natural England (2010) *Higher Level Stewardship. Farm Environment Plan (FEP) Manual: Technical Guidance on the Completion of the FEP and Identification, Condition Assessment and Recording of HLS FEP Features*, (NE264) (3rd ed.). Available at: https://webarchive.nationalarchives.gov.uk/20140605151020/http://publications.natu ralengland.org.uk/publication/32037.

Robertson, M. (2006) 'Nature That Capital Can See: Science, State and Market in the Commodification of Ecosystem Services', *Environment and Planning D: Society and Space*, 24(3), pp. 367–387.

Sandbrook, C.G., Fisher, J.A. and Vira, B. (2013) 'What Do Conservationists Think About Markets?' *Geoforum*, 50, pp. 232–240.

Sullivan, S. (2010) ' "Ecosystem Service Commodities" – A New Imperial Ecology? Implications for Animist Immanent Ecologies, With Deleuze and Guattari', *New Formations*, 69(18), pp. 111–128.

Sullivan, S. (2012) 'Financialisation, Biodiversity, Equity: Some Currents and Concerns', *Environment and Development Series*, (16), Malaysia: Third World Network.

Thomas, N. (1992s) *Entangled objects*. Cambridge, MA: Harvard University Press.

Yusoff, K. (2011) 'The Valuation of Nature: The Natural Choice White Paper', *Radical Philosophy*, 170(Nov/Dec), pp. 2–7.

Part II
Spacing value

7 Regimes of value in a Chicago market

Tim Cresswell

Places are made from things: things and practices and meanings (Cresswell, 2014). In the language of phenomenology, places gather (Casey, 1996). In the terms of assemblage theory, they assemble (DeLanda, 2006; Dovey, 2010). Whatever the theoretical language, one way into place is through the things that gather there. Places are also sites of value. Forms of valuing (and devaluing) help to distinguish a rich sense of place from mere location. The place I focus on here is the area around the Maxwell Street Market in the near west side of Chicago. This chapter is part of a wider project exploring a hundred years of this market and the area around it through various practices of valuing, including writing, photographing, archiving and planning. The project is, simultaneously, about the things of Maxwell Street and the practices that value and devalue those things. The concerns of this chapter thus reflect those of the wider project (see Cresswell, 2019).

Maxwell Street Market was the largest open-air market in North America for much of the last century. Jewish street pedlars started it in the 1880s, and by the middle of the 20th century, it was largely associated with the African American population and the development of the Chicago blues. In the last two decades of the 20th century, it was gradually erased by gentrification processes led by the University of Illinois in Chicago (Cresswell, 2012; Cresswell and Hoskins, 2008; Grove, 2002; Berkow, 1977). Throughout its history, the market and the area around it was a site of heterogeneous gathering and assemblage of things, practices and stories. Here I provide just a sliver of this place through what, on the face of it, are highly disparate stories. The first half of the chapter focuses on two objects as they appear in a variety of archives – hubcaps and DIY musical instruments called stradizookys. The second half of the chapter widens the scope of investigation and considers the role of value in the process of declaring the area a tax increment financing (TIF) district in 1999.

First, though, a word on value. Exploring a market place means returning to one of the root meanings of the city – a place where exchange happens (Weber, 1960; Jacobs, 1969). From the writings of Max Weber to the heretical theory of urban origins proposed by Jane Jacobs to contemporary work in the Marxist political economy tradition, the city is a site characterised by, even originating in, the creation of surplus value through trade. Exploring a market as a place, then, means exploring the most urban of urban sites.

The processes of valuing and exchange at Maxwell Street were heterogeneous and multi-scalar. The Maxwell Street Market was (and still is – in a relocated form) a flea market. It was a place where a significant portion of what was for sale was second-hand. Shoppers came to the market in large numbers expecting to get bargains. At the same time, the stallholders expected to, in a telling term, "cheat you fair". The process that ensued was bargaining. This was a practice of valuing that led to (in some instances) the continuing biography of an object as it moved from Maxwell Street to the domestic spaces of shoppers from across Chicagoland. Another form of valuing was the way Maxwell Street itself has been consistently valued as a destination – a place to go – a stop on a tourist itinerary and a place for a bargain. The visits of photographers to the market over a hundred years reveal a kind of bourgeois valuing of the place of the other – a place to slum it for a day. It was a place to find surreal juxtaposition, picturesque poverty and constant drama. Different but related forms of valuing can be seen in the works of novelists and sociologists and, indeed, in my own fascination with the place (Motley, 1947; Wirth, 1928).

Towards the end of its life – in the last three decades of the 20th century – Maxwell Street was the site of heated debates about the gentrification process that was gathering pace and would eventually lead to the demise of the market. Many of the arguments that swirled around Maxwell Street were arguments about the values of things. Briefly put, discussion centred on whether certain objects in the Maxwell Street area, and the area itself, deserved to persist or be discarded. The idea of "regimes of value", derived from the work of Appadurai, suggests certain contexts in which things are ascribed value (Jamieson, 1999; Schlosser, 2013; Appadurai, 1986). It performs a critique of the idea of inherent value at the same time as it dispenses with the differentiation between commodity value and gift value (as two subsets of exchange value). Things travel through these regimes and in doing so have "careers" or "biographies" (Kopytoff, 1986). In Appadurai's terms, they have "social lives" (Appadurai, 1986). In this sense, the objects of Maxwell Street, and Maxwell Street itself, are fluid concretisations of the relations between the human and the nonhuman worlds – of the way value is ascribed to objects.

In the end, Maxwell Street became the site of a protracted decade-long struggle over valuing as the University of Illinois, Chicago, in a classic act of gentrification, bought up the land and transformed it into an expensive "university village" built along trendy new-urbanism principles. Fighting against this process, protesters looked to the regime of heritage to save the remaining landscape as an official historic place. They lost (Cresswell and Hoskins, 2008). More recently, the area has been declared a TIF district so that private developers can use future taxes to fund the development of the area for 22 years. During this period, the developers would benefit from the taxes resulting from the increasing property values of the area, while these same funds could not be used for publicly funded improvements to civic amenities. In order for this place to be classified in this way, it had to be declared essentially valueless – as decrepit, run down and in need of renewal. As we shall see, the process of TIF is essentially a magical process of diagnosing valuelessness in the landscape and then making that valuelessness valuable. It is a way of inserting the micro-geographies of a run-down landscape into the macro-geographies of the global financial system.

To value something is to include it in some way in a world of significance. To value something is to decide it is worthy of inclusion in a sphere that is itself deemed worthy. Valuing is an act of inclusion and exclusion. I am thinking here of "value" as a verb more than as a noun: less the idea of the worth in things and more the idea of making things worthy.

My exploration of Maxwell Street, then, has the notion of valuing at its heart. It is about a place where forms of negotiation over value were central to its existence. It was full of things that were constantly leaving and entering differing and sometimes competing regimes of value. It was a place that was, itself, valued in different ways by shoppers, tourists, writers, photographers and urban planners, among others. Finally, it was a place that was erased following a contracted battle over how, exactly, the place should be valued. In the end, through TIF, its supposed lack of value was used as a way of entering the place into global circuits of securitisation and financialisation. With this in mind, let's proceed to some of the things that were valued as they entered and left Maxwell Street.

Hubcaps

In 1974, a reporter discovered 72-year-old Leamon Reynolds next to a six-foot high pile of hubcaps selling for around $1.50. Reynolds, it turns out, "can find you a 1952 Ford hubcap in five seconds, thanks to his secret filing system. Where does Reynolds get his fantastic stock? From state road workers, he says, they pick them up while patrolling expressways" (Star, 1974).

Six years earlier, Reynolds's hubcap stand had attracted the attention of photographer James Newberry, who was entranced enough with its silvery stock to take a picture that can be found today in the archives of the Chicago History Museum (Figure 7.1). The picture can also be seen on the remaining block of Maxwell Street, on the side of one of the mock piles of boxes that remind the present-day visitor of the place this once was (Figure 7.2). In this way, the humble hubcap found its way first into the official archive of the market and then back, in simulated form, onto the street where the market once was.

Newberry's image of hubcaps and brooms is like a still life. There are no people. Many of the photographs in the archive are of intense crowds of peoples and varieties of performance that take place in a market. This, on the other hand, appears as an accidental arrangement of confused forms and surfaces. The beauty of the photograph lies in the configuration of the hubcaps as discs and the straight lines of the brooms. It presents us with aesthetically pleasing confusion.

Photography is one way that hubcaps in Maxwell Street entered regimes of value. But there are other ways. The hubcap appears repeatedly in the words of those who argued for the demolition and relocation of the market. Its banal materiality became a vehicle for a discourse that framed the market as a site of dubious moral order. In the archive of the University of Illinois in Chicago are a series of letters written to Mayor Richard J. Daly's office supporting the relocation of

Figure 7.1 Hubcaps, Union Street, north of Maxwell Street; Chicago (ill), 1966; Photographer –
James Newberry;

Source: Photograph – ICHi-20332, Chicago History Museum

Figure 7.2 Hubcaps on a mock pile of crates, sculpture on Maxwell Street, 2009

Source: Photo by author

the market. One, from Michael Shea of "Buy a Tux" Formal Wear Superstore on nearby West Roosevelt Avenue reads,

> Where do the goods come from? On more than one occasion, we bought my own hubcaps on Maxwell Street (15 minutes after they were stolen off our car). The absence of this can only have a positive effect on the area and Chicago proper.[1]

The hubcap was linked to much more serious pronouncements of moral dissolution in a letter from the University's head gymnastics coach on 3 November 1993.

> When I think of Maxwell Street, I think of three things:
>
> 1   Garbage
> 2   Crime
> 3   Perversion
>
> . . .
>
> In regard to crime, I personally have witnessed drug deals, prostitution, car thefts, and creeps prowling the area daily. I have to buy back my own hubcaps, radio and accessories two or three times a year.[2]

The story of finding your own hubcaps at Maxwell Street just after they have been stolen is one of the most often-told stories of Maxwell Street. It is told so often that, in most cases, it is unlikely to be true. Who, after all, knows what their own hubcaps look like? This is the way a place becomes storied. A story is told over and over until it sticks – until it becomes so much common sense.

Hubcaps clearly exerted an influence on Maxwell Street. To photographers such as James Newberry, they presented an aesthetic opportunity. Piled up in profusion, they created form and contrast – they became a sign of the object richness of the market. They were beautiful. To others, less enamoured of the market, they represented an amoral place in which hubcaps were signs of a wider broken society – linked to crime and perversion.

Stradizooky

In the archives of the writer Ira Berkow, the author of *Life in a Bazaar*, there are the transcripts of all the oral histories that he collected for the remarkable book – an account of Maxwell Street collated from those who lived and worked there over the years (Berkow, 1977). Included in the archives are some transcripts that never made it into the final manuscript. These include an interview with Tyner White – an interview that hilariously goes nowhere:

> IB What does the street mean to you? What does Maxwell Street mean to you?
>   TW What Maxwell Street means to me? Essentially, it's something that the city means. The city means an exchange market. You visit there, and you

offer others things you don't need, and you get things from them that they don't need. These are wares. Wares are things which were. And now I don't need it anymore.[3]

In the home of economics professor Steve Balkin, an advocate for the market, I noticed some curious wooden objects hanging from shelves. These, he told me, are Stradizookys – musical instruments made from scrap bits of wood and other junk. Tyner White made them. The word "Stradizooky" is derived from Stradivarius, the renowned violinmaker and Suzuki, the originator of the Suzuki method of teaching children to play violin. The Stradizooky combines a passion for recycling wood with a quest for racial/ethnic togetherness. One example in Balkin's loft has "Blacks + Jews = Blues" inscribed on it. Tyner White graduated from the master of fine arts programme in creative writing at the University of Iowa. After a flirtation with poetry, he dedicated himself to educating people about the wonders of wood and the necessity of creative recycling. Like many before him, he gleaned stuff from the Maxwell Street area to work on his new inventions. A journalist from the Chicago *Reader* was impressed with White's creativity:

He's built a mad hatter's assortment of prototypes: a possibly functional tape dispenser in the shape of a cat, rubber-tipped walking sticks with handles of telephone wire, oversize sculptural chess pieces sporting shiny metal screws for arms, a deeply discordant toy violin. "Here," he says, offering a box of lumber scraps that he's sanded and bevelled. "Take a diamond."[4]

Tyner White was a central figure in the Maxworks artists collective, who inhabited 716 Maxwell Street until they were forcefully evicted to make way for the University Village in March 2002. Theirs was the last inhabited building on the old street. Once evicted, White took his gleaning project to the Resource Centre, where he founded the Maxwood Institute of Treeconomics, which sits alongside the Creative Reuse Warehouse – a place where artists can get scrap materials cheaply for the construction of installations and other artworks.

On New Year's Eve at the turn of the millennium, an old Nabisco factory at 720–724 West Maxwell Street mysteriously went up in flames. Tyner White and the residents of the Maxworks collective witnessed the fire. Arson was suspected. White recounted his experience to a journalist from the *Chicago Reader*:

"It's like a war and they're trying to exterminate our resources," says White, who has built thousands of bizarre instruments and knickknacks out of "recycled" scrap wood, including the "Stradizooky," a violin-like musical instrument, and his trademark "Toker," a device for smoking marijuana that he claims will help replace the demand for cigarettes. He says he now hopes to "get a moratorium on bulldozing" and to pave Maxwell Street with bricks salvaged from the demolished factory, which, he says, had also contained remnants from the days when the market was predominantly populated by central European Jews.[5]

White's recollection of the burning building is linked to his use of recycled wood to suggest an alternative kind of valuation of the place of Maxwell Street and the things in and around it: a valuation based on reuse and recycling rather than destruction and demolition. When I visited the relocated Maxwell Street Market at its new site on South Desplaines Street, on its hundredth birthday, there was Tyner White playing away on a Stradizooky as the Maxwell Street Blues Band did its thing.

The Stradizooky, like the hubcap, is valued in particular ways that are connected to the place it is associated with: Maxwell Street. It tells us about Tyner White's valuation of things that others consider to be junk and evidence of the decay of the Maxwell Street area. A piece of wood becomes a "diamond" or a musical instrument. This particular form of valuation is most evident in another of White's appearances in the distributed archive. He turns up in a City of Chicago Community Development Commission Meeting Report for a meeting held on 26 October 1993 to consider the future of the market. He pointed out that the University of Illinois in Chicago had a terrible recycling record and offered to take on some of the work at the Maxworks Institute (still on Maxwell Street at the time):

> We could convert some of the scrap lumber into workroom shelves and other kinds of things for the physical plant,
> And I would like to mention that in our block are several shuttered buildings which the University acquired over the years, in which they have manifested a wish to tear down.
> Now the reason is that ten years from now, it would then be possible to install a four or ten or forty-six million research building.
> . . .
> I would recommend that the University consider recycling the warehouse buildings on Maxwell Street, make them available for use in a joint venture and find out how much this University can contribute to solving the recycling crisis.[6]

White's advice was ignored.

Tax increment financing

Through the examples of hubcaps and the Stradizooky, we have seen how things, at a micro-geographic scale, enter and leave regimes of value in a particular geographic context – that of Maxwell Street. They are ingredients in the gathering of things that was Maxwell Street. In the remainder of this chapter, I focus on one particular process that Maxwell Street as a whole entered into. We have seen how both hubcaps and the Stradizooky entered into debates about the value of Maxwell Street as a whole. Such things, as signs of decay and blight, also played a role in a macro-geographic process of valuing and devaluing called TIF.

The fluid, restless nature of capital constantly comes up against the friction of fixed capital – the relative intransigence of bricks and mortar (Harvey, 1982). One way to navigate this problem and reduce the friction is to create financial instruments that bundle and abstract the idea of "place value" and make it transferable. It is this process that I want to explore now through another set of (de)valuing practices that were applied to the area around Maxwell Street Market in the late 1990s.

In 1999 the City of Chicago decided to designate the area around Maxwell Street as the Roosevelt/Union Redevelopment Project Area under the Tax Increment Financing Program. TIF works by allocating yet-unrealised increases in property taxes from areas that are approved as TIF zones to pay for "improvements" in that area. The idea was not native to Chicago; it had been pioneered in California as early as 1952. In all, 48 states have embraced it as a means of financing improvements in infrastructure and economic development. It had become possible to raise money in this way through an Illinois state law in 1977, but the first Chicago TIF district – the Central Loop – had not come into existence until 1984. Since then, the City of Chicago has used TIF financing in over 100 areas of the city and has been its most enthusiastic advocate (McGreal et al., 2002; Weber, 2010; Gibson, 2003).

This is how TIF works. The first step is to designate a neighbourhood as "blighted" – as devalued or relatively valueless. Once the designation of a blighted area has taken place, an amount is allocated as the tax baseline based on tax revenues at the point of designation. This is then frozen and for the following 23 years, there is no additional revenue from this tax base available for local school districts, roads, parks or any other general civic amenity. Any additional tax revenues that are then collected in subsequent years are used to finance various forms of development until the TIF district definition comes to an end after a 23-year period. All new tax revenues go to neighbourhood (local) public works or subsidies to private developers. These developers can begin work based on future tax revenues. This process favours big developments on large parcels of land where large increases in tax revenue can be quickly realised. As with the urban renewal programmes of an earlier period, the TIF process uses eminent domain (compulsory purchase by the state) to purchase land. Unlike urban renewal programmes, the proceeds almost always flow directly to private property developers with little transparency or public oversight. Many argue that this diverts money away from public bodies that would have benefited from increasing tax revenues had TIF not been implemented.

Once the TIF district has been dissolved, the city benefits from the increased tax base, which results from the development process. The idea is that as blighted properties are improved through the investment of TIF dollars, their value will increase along with the newly generated revenue. The difference between the tax base line and the new, more-valuable property is captured and reinvested for improvements, thus further increasing the values.

Obviously, there is a catch here as the increased tax revenues are not available in advance. The local state therefore has to invent a financial instrument to provide funds up front. To do this, it issues bonds with future tax revenues as security. The bonds are sold through negotiated sales to a variety of investors, including pension funds, across the world. "In this way, cities obtain capital by turning the rights to their own heterogeneous property tax base into standardized tradable assets – often without the knowledge of the individual property owners paying the tax bills" (Weber, 2010: 258). This puts the landscape into a complicated and fragile network of risk that is spread across a vast and unstable system. The landscape is enrolled into a topology of risk that it was previously outside of. The city government cannot assure the increase in tax revenues that the investment is based on – the main drivers of property value are also situated well beyond the TIF zone or even the city of Chicago.

TIF designation is part of what Leyshon and Thrift have called the "capitalization of almost everything". They have argued that the current round of financial capital investment is based on an innovative reconfiguration of securitisation:

> we are in a period in which the process of securitization, which has driven so much of international finance since the 1980s, is engaged in a fresh round of tracing value to its source – or rather sources – since what we can see now is an impulse to identify almost everything that might provide a stable source of income, on which more speculation might be built, being brought into play.
> (Leyshon and Thrift, 2007: 98)

Securitisation refers to how a loan is secured. The most obvious example is how a mortgage is secured by the value of the property being bought. If you cannot pay your mortgage, then the bank has your house to recoup its losses. Leyshon and Thrift's argument is that this process is increasingly returning to previously moribund elements of the physical landscape in order to find new forms of security that will enable the wilder processes of speculation that have been identified as the frontier of contemporary financial capitalism. These moribund elements of our everyday worlds that are being used for security include infrastructure such as roads, parks and water systems – previously thought of as the least exciting components of the financial landscape.

These previously overlooked elements of geography are being inserted into the financial system through the innovative bundling of assets produced through computer software. This then makes it possible to sell bonds with new kinds of security, and these bonds become part of a global system of pension funds and other kinds of bundled investments designed by financial industry specialists who are seeking to diversify assets and risks.

The Roosevelt/Union area surrounding Maxwell Street was made a TIF district on 21 May 1999 and will cease to be one on 21 May 2022. At the time of its inception, it was one of around 70 TIF districts, most of which had been approved in the two years immediately preceding it. In 1999, around 7.7% of all property in Chicago was in a TIF zone (Schwartz, 1999). As with all TIF zones, the Roosevelt/Union area underwent an eligibility study to ensure that the district counted as "blighted". The study was hired out to a consultant – Louik Schneider and Associates, Inc – who then submitted an eligibility plan to the city of Chicago's Community Development Commission (CDC) (who have never turned down an application). The commission were then obliged to order a public hearing (which involved telling only property owners in the area, 14 days before the hearing). The public hearing had no legal standing, and the city could do what it wanted after the CDC vote. The process then passed through the finance committee and the city council. Following this process, the 58 acres of the Roosevelt/Union area, with a 1997 property value of $3,987,742, became a TIF district.

The Roosevelt-Union Redevelopment Plan and Project report was published in October 1998. The objective of the plan, it stated, was to "encourage mixed-use development, including new residential, institutional and commercial development within the area" and to enhance the city's tax base and preserve the values of existing property. It stated that the area was well suited to mixed use due to the close proximity of transport infrastructure, including the Chicago Transport Authority bus and train lines and major highways (Chicago, 1998).

The area, the report states, was formally designated a slum and blighted area on 11 August 1966 (as part of the earlier and discredited round of "urban renewal") and can therefore be a TIF. In addition, it cites a number of other existing city goals and policies that would support the TIF designation, including the University of Illinois Master Plan of 1977 and the 1996 Chicago Zoning Ordinance.

These general details are interspersed with "design objectives" for the improved area, including "high standards of appearance" and the need to encourage

> a variety of streetscape amenities which include such items as sidewalk planters, flower boxes, plazas, variety of tree species and wrought-iron fences where appropriate.
>
> (Chicago, 1998, p. 9)

To be considered "blighted", an area had to have five or more "factors" that, combined, would make the area "detrimental to the public safety, health, morals, or welfare" (Chicago, 1998, p. 10). It also had to be the case that the area would have little chance of being "developed" without action from the city. The following was the list of factors that contribute to blight:

> Age; dilapidation; obsolescence; deterioration; illegal use of individual structures; presence of structures below minimum code standards; excessive vacancies; overcrowding of structures and community facilities; lack of ventilation, light or sanitary facilities; inadequate utilities; excessive land coverage; deleterious land use of layout; depreciation of physical maintenance; or lack of community planning.
>
> (Chicago, 1998, p. 10)

Nine of these were discovered in the area, including, to a major extent, age, dilapidation, deterioration, excessive vacancies, excessive land coverage and depreciation of physical maintenance. Each of these is then defined and mapped. The following is the definition of "age" as a factor in blight:

> Age presumes the existence of problems or limiting conditions resulting from normal and continuous use of structures which are at least thirty-five

(35) years old. In the Redevelopment Project Area, age is present to a major extent in sixty-seven (67) of the seventy-three (73) (ninety-one and seven-tenths per cent (91.7%) buildings and in thirteen (13) of the sixteen (16) (eighty-one and three-tenths per cent (81.3%) blocks in the Redevelopment Project Area.

(Chicago, 1998, p. 12)

The report concludes that the redevelopment of the area will cost $103,000,000. A base line value for all the property (EAV: equalized assessed value) was set at $3,968,563, with an expected EAV for 2008 of $48,000,000 to $55,000,000.

Conclusion

Maxwell Street in the 1980s was the kind of place described in the following terms:

> During the week, the dusty vacant lots are more desolate than ever. Shabbily dressed old men sit silently on crumbling stoops and drink wine in garbage-strewn alleys; the few remaining buildings sag wearily, burned out stairwells and boarded-up windows telling the perennial urban story of neglect and decay. ("The Sunday morning market may be in danger, but thanks to a new generation of bluesmen the music is as strong as ever".)
>
> (Whiteis, 1988, p. 8)

And consider Maxwell Street now. The remaining two blocks feature a pastiche of facades saved from demolition and used to provide the fronts for shops, restaurants and a parking lot. It is part of an area described as University Village – a prestigious area of mid to high-end town houses and apartments built on neo-traditional town-planning principles. A pamphlet mailed to employees of the nearby University of Illinois, Chicago, in 2001 advertised the area in the following way:

> "Yesterday's Heritage. Tomorrow's Treasure".
> Chicago's newest, most convenient, most thoughtfully planned neighbourhood . . . a great life in the city.
> University Village presents traditional Chicago-style architecture on tree-lined streets.
> Townhome exteriors feature varying rooflines.
> The site plan features neighborhood parks and green space corridors.
> "Chicago's next great neighborhood".

The stories of value recounted here describe moments in this transformation. They reveal how it is necessary to connect the materiality of place (by which I mean, in this instance, things like hubcaps, wood and wrought-iron fencing), to meanings and narratives of place (such as the designation of "blight"). The combination of things and value and the regimes that authorise these combinations are key to deciding whether a place fades or endures.

Discussions of hubcaps, scrap wood and a host of other "things" in and around Maxwell Street form part of a wider evaluation of the place that was Maxwell

Street. Similarly, the act of defining an area through maps and statistics during the TIF process forms part of the quasi-scientific process of devaluing an area that is a necessary precursor to the revaluing process. The designation of "age" to property contributes to a definition of obsolescence: the notion that something that is no longer of the times is thus out of date. This designation ties a narrative of lack of value into the material landscape to legitimate certain practices of demolition and redevelopment that suit the purposes of private capital. Narrative becomes part of the landscape. This place with its hubcaps and scraps of wood (among many other objects) is defined as decrepit, decayed and obsolete. This lack of value – or negative value – is then repackaged as a new form of value that can produce a new kind of place: one without the Maxwell Street Market.

Notes

1 Letter dated 30 October 1993 to Richard J Daly's office from Michael Shea of Buy a Tux Formal Wear Superstore on W. Roosevelt Avenue, UIC University Archives, Associate Chancellor, South Campus Development Records 003/02/02. Series 1, Box 11 Folder 11–90.
2 Letter dated 3 November 1993 to Richard J Daly's office from C.J. Johnson, UIC heard gymnastics coach, UIC University Archives, Associate Chancellor, South Campus Development Records 003/02/02, Series 1, Box 11 Folder 11–93.
3 Ira Berkow Archives – Special Collections – University of Illinois, Chicago, Tyner White 78–16 Box 1, File 1–8, Page 8.
4 Deanna Isaacs, "The Collected and the Ultimate Collector; Tyner White's Glorious, Homeless Hoard" 7 April 2005: www.chicagoreader.com/chicago/the-collected-and-the-ultimate-collector-tyner-whites-glorious-homeless-hoard/Content?oid=918451 (Accessed 21 October 2012)
5 Kari Lyderson, "Burnt Out". Will the New Year's Eve fire break the spirit of preservationists fighting to save what's left of Maxwell Street?' January 6, 2000: www.chicago reader.com/chicago/burnt-out/Content?oid=901112 (Accessed 21 October 2012)
6 City of Chicago Community Development Commission Meeting Report. Public meeting held at 7.00 p.m. on 26 October 1993 at YMCA, 1001 W Roosevelt Rd. UIC University Archives, Associate Chancellor, South Campus Development Records, 003/02/02. Series X Box 51 File 432, pp. 118–122.

References

Appadurai, A. (1986) 'Introduction: Commodities and the Politics of Value', in Appadurai, A. (ed) *The Social Life of Things: Commodities in Cultural Perspective*. Cambridge, MA: Cambridge University Press, pp. 3–63.
Berkow, I. (1977) *Maxwell Street: Survival in a Bazaar*. Garden City, NY: Doubleday.
Casey, E. (1996) 'How to Get from Space to Place in a Fairly Short Stretch of Time', in Feld S and Baso K (eds) *Senses of Place*. Santa Fe: School of American Research, pp. 14–51.
City of Chicago (1998) 'Roosevelt-Union Redevelopment Plan and Project', Prepared by Louik/Schneider & Associates, Chicago: City of Chicago. Available https://www.chicago.gov/content/dam/city/depts/dcd/tif/plans/T_068_RooseveltUnionRDP.pdf.
Cresswell, T. (2012) 'Value, Gleaning and the Archive at Maxwell Street, Chicago', *Transactions of the Institute of British Geographers*, 37(1), pp. 164–176.

Cresswell, T. (2014) 'Place', in Lee R, Castree N, Kitchin R, et al. (eds) *The SAGE Handbook of Human Geography*. London: Sage, pp. 7–25.

Cresswell, T. (2019) *Maxwell Street: Writing and Thinking Place*. Chicago: University of Chicago Press.

Cresswell, T. and Hoskins, G. (2008) 'Place, Persistence and Practice: Evaluating Historical Significance at Angel Island, San Francisco and Maxwell Street, Chicago', *Annals of the Association of American Geographers*, 98(2), pp. 392–413.

DeLanda, M. (2006) *A New Philosophy of Society: Assemblage Theory and Social Complexity*. London, New York: Continuum.

Dovey, K. (2010) *Becoming Places: Urbanism/Architecture/Identity/Power*. London: Routledge.

Gibson, D. (2003) 'Neighborhood Characteristics and the Targeting of Tax Increment Financing in Chicago', *Journal of Urban Economics*, 54(2), pp. 309–327.

Grove, L. (2002) *Chicago's Maxwell Street*. Chicago, IL: Aracadia Pub.

Harvey, D. (1982) *The Limits to Capital*. Oxford: Blackwell.

Jacobs, J. (1969) *The Economy of Cities*. London: Jonathan Cape.

Jamieson, M. (1999) 'The Place of Counterfeits in Regimes of Value: An Anthropological Approach', *Journal of the Royal Anthropological Institute*, 5(1), pp. 1–11.

Kopytoff, I. (1986) 'The Cultural Biography of Things: Commoditization as Process', in Appadurai A (ed) *The Social Life of Things: Commodities in Cultural Perspective*. Cambridge, MA: Cambridge University Press, pp. 64–94.

Leyshon, A. and Thrift, N. (2007) 'The Capitalization of Almost Everything – The Future of Finance and Capitalism', *Theory Culture & Society*, 24(7–8), pp. 97–115.

McGreal, S, Berry, J, Lloyd, G. and et al. (2002) 'Tax-Based Mechanisms in Urban Regeneration: Dublin and Chicago Models', *Urban Studies*, 39(10), pp. 1819–1831.

Motley, W. (1947) *Knock on Any Door*. New York, London: D. Appleton-Century Company.

Schlosser, K. (2013) 'Regimes of Ethical Value? Landscape, Race and Representation in the Canadian Diamond Industry', *Antipode*, 45(1), pp. 161–179.

Schwartz, C. (1999) *Chicago Tif Encyclopedia*. Chicago: Neighborhood Capital Budget Group.

Star, J. (1974) 'Maxwell Street Lives', *Chicago Tribune Magazine*. Chicago: Neighborhood Capital Budget Group, p. 1.

Weber, M. (1960) *The City*. Heinemann.

Weber, R. (2010) 'Selling London: City Futures: The Financialization of Urban Redevelopment Policy', *Economic Geography*, 86(3), pp. 251–274.

Whiteis, D. (1988) 'The Sunday Morning Market May Be in Danger, But Thanks to a New Generation of Bluesmen the Music Is as Strong as Ever', *The Reader*. Chicago: Neighborhood Capital Budget Group, p. 8.

Wirth, L. (1928) *The Ghetto*. Chicago, IL: The University of Chicago press.

8 Urban-planning practice and the transformation of value in China

Evidence from the city of Yangzhou

Xu Huang, Jan van Weesep and Martin Dijst

Introduction

Ancient Chinese society was a unified feudal state with deep roots in Confucianism. It encouraged the dominant position of centralized political power and related social order (Shek, 2006). Influenced by Confucianism, China's urban-planning practice started three thousand years ago and emphasized the principle of feudal hierarchy – where urban constructions served the king (Wu, 1986). The establishment of a socialist regime in 1949 and the adoption of the 1978 reform introduced socialist values and modern thought (Shek, 2006). China's Socialists considered the city as a production centre, symbolizing the advanced productive forces required to serve the planned economy (Yeh and Wu, 1998). The 1978 reform placed more concern on people's living environment in urban-planning practice (Ma, 2004).

These transformations represent changes in the value system that redefined the meaning of the city and the role of urban planning. Specifically, in this chapter, we conceive values as the prevailing norms, beliefs, imperatives, guiding principles, logics and priorities of society that shift across place and time. Such values are acted on and can shape physical urban form (Castells, 1983). This chapter traces the transformation of the meaning of "city" in the Chinese context of changing value systems and assesses its impacts on urban-planning practice, including urban design, architecture, layout, regulatory principles of transport, land use and so on. We pose the following questions: what transformation in the meaning of the city was brought by the change of the value system, and how does China's urban-planning practice reflect this transformation?

This chapter employs Yangzhou city as a case study: it is a state-level historic cultural city in Jiangsu province with a history of approximately 2,500 years of urban development. Its development can be divided into three phases: imperial China, socialist China before the 1978 reform and transitional China after the 1978 reform.

Locating Yangzhou city

Yangzhou city is located at the confluence of the Yangtze River and the Beijing-Hangzhou Great Canal. The city has acted as a canal hub, connecting North China and South China. Because of its specific geographical location, almost every dynasty valued its national strategic importance and invested in its urban infrastructure. With the rich material legacies of each subsequent regime, it provides an excellent case study to examine the relationship between prevailing value systems and urban-planning practice in China's medium-size cities.

Currently, the total area of the municipality is 6,591 km^2 (Yangzhou city has an 82 km^2 built-up urban area), and the Yangtze river flows through its southern section. There are about 50 small rivers and three big lakes within its territory. The most recent census, in 2010, shows the municipal population has 4.59 million people, with 1.22 million living in central Yangzhou city (Yangzhou municipal government, 2011).

Urban construction serving the king in imperial China

Although, the concept of modern urban planning was introduced to China from Western nations in the beginning of 20th century, urban planning in China has its origins three thousand years ago, in the Zhou Dynasty (1046–256 BCE) (Zhang, 2002). At that time, the cosmic view *"tianyuan difang"* [the sky is circular, while the land is square] was widely accepted (Smith, 2007). Confucianism adopted this idea and thereby defined the meaning and layout of the city. A square place symbolized the land in that cosmic model, and the central space was left for the exclusive use by the king to show his dominant position and power.

According to that framework, the classical handbook of urban planning *"Kao Gong Ji"* [The Craftsmen's Record in Zhou Dynasty] established the principles of the ideal city and expressed formal design guidelines that included orientation, enclosure, shape, arteries and hierarchy. In the handbook, the spatial form of the city and its architecture reflected the caste system of social structure at that time, stipulating different sizes for the city of the emperor and the cities of the feudal lords and princes. The area of each city, the height of its walls and the width of the roads were scaled according to social rank (Wu, 1986). More specifically, *Kao Gong Ji* says that *"jiangren yingguo, fangjiuli, pangsanmen"*: the standard urban spatial form should be a square (about 1.5 km by 1.5 km) protected by a city wall and moat at its boundary; the royal palace should be located at the centre; and offices and housing of government officials should be sized according to the user's position in the social hierarchy.

These urban conventions were favoured by the dominant classes, the royal family and related feudal lords. Based on Confucian guidance of urban-planning practice, the dominant class had built cities at the geographical centre of regions, to supervise peasants in rural areas and to maintain the central dominant position. Following that idea, Yangzhou city was first built by King Wu in the Zhou Dynasty (around 486 BCE), acting as a military castle. The layout of the city was designed to meet the criteria and values specified in *Kao Gong Ji* (Wang, 2012), reflecting the meaning of the city under Confucianism principles.

The Sui Dynasty (581–618 CE) started to build the Beijing-Hangzhou Great Canal, and Yangzhou city became a commercial hub on the canal. Emperor Yang (604–618 CE) visited Yangzhou city while boating along the canal on three occasions between 605 CE and 618 CE (see Yangzhou municipal government, 2012). Yangzhou city was therefore carefully rebuilt to meet the needs of the royal family under the Confucian guidance of an ideal city. There were a lot of luxurious royal gardens, and the Beijing-Hangzhou Great Canal acted as the moat around the city wall. At this time, the old city covered 5 km^2, with a width of 2.0 km and a length of 2.5 km.

Its development boom lasted into the Tang Dynasty (618–907 CE). Owing to its canal infrastructure, Yangzhou city became the economic centre in Southeast China. The main commerce was the transportation of salt and silk – moving salt from the north to the south and moving silk from the south to the north (see Yangzhou municipal government, 2012). However, during the Song Dynasty (Northern Song, 960–1127 CE; Southern Song, 1127–1279 CE), the development of Yangzhou city dramatically decreased because the canal transportation was seriously interrupted by the war between the Song Dynasty and northern minority regimes (kingdoms Liao, Xia, Jin and Yuan). Later, after periods of rise and decline, the city regained prosperity during the Ming Dynasty (1368–1644 CE) (Ji, 2002).

The city developed again during the reigns of Emperor Kangxi (1661–1722 CE), Emperor Yongzheng (1722–1735 CE) and Emperor Qianlong (1735–1796 CE) in the Qing Dynasties (1644–1912 CE), due to the further development of the salt industry and the silk industry. For instance, tax from the salt industry would contribute almost half of the revenue of the national government at that time. As the commercial centre of the salt market, merchants from all parts of China gathered in Yangzhou city. Therefore, it quickly developed into a tourist and consumption centre in Southeast China (Ho, 1954). These merchants also invested a lot in urban constructions. They built private gardens in the city and put their head offices there; they dug small canals to connect their gardens to the big canal. They competed to show their wealth via these constructions – for example, whose garden is biggest and most beautiful (Wang, 2012). In this sense, the Confucian principles mixed with the value systems of the merchants, who were high in the social hierarchy at this time, and this combination of values shaped the city.

After the First Opium War, in 1840, the national government gradually opened sea ports and sea transportation, and it permitted the construction of railways by foreign enterprises. Canal transportation lost the dominant position, which affected the development of Yangzhou city. In contrast, Wuxi city, where silk was produced, developed a manufacturing industry instead of transportation and commercial industry, and it quickly became the regional centre in Southeast China (see Yangzhou municipal government, 2012). The old downtown of Yangzhou city consequently fell into disrepair, and most public services were not functioning, due to damage wrought by the wars. A large number of the early royal gardens and palaces disappeared. Currently, the inner-city area still has over 500 clusters of historic buildings and 148 sites of cultural heritage that allow the meaning of city under the Confucianism principles to be traced (see Yangzhou municipal government, 2012).

Urban planning as a tool of a planned economy in socialist China

After the Communist Revolution in 1949 the newly established government seized the opportunity to rebuild the cities as models of socialist values. China's socialists held the view that the city should be the place where the most revolutionary people, the "working class", live. The city should act as a production centre, symbolizing the advanced productive forces (Yeh and Wu, 1998). This meaning of the city was consistent with the practical needs of the party – since it intended to develop military-related heavy industries to protect the newly built Socialist regime. in particular, the party needed to develop urban areas to locate 156 industrial projects aided by the Soviet Union and a considerable number of other supplementary local projects (Zhang, 2002). With the adoption of Soviet-style orthodox planning commissions, central and local governments designed comprehensive economic (five year) plans to prioritize industrial military production (Xie and Costa, 1993).

Urban planning was subject to the strategy of economic planning commissions and became a tool to realize socialist development (Yeh and Wu, 1998). Urban planners were tasked to coordinate construction projects, select factory location and plant sites, design the layout of industrial towns, divide functional zones and arrange service facilities (Zhang, 2002). However, in the territory of the production site, each state-owned enterprise (work unit in English, *danwei* in Chinese), rather than the municipal government, took charge of construction. This delegation of responsibility employed the principle of self-containment: the residents could live and work inside the site. Thus, a special urban component, the gated *danwei*, was created. It provided not only a workplace but also employee housing, health care, food distribution and other basic social services (Xie and Costa, 1993).

No matter how large or small, each *danwei* was surrounded by an enclosed wall. Inside the wall, land use was divided by various functions, such as workshop area, office area and residential area. Larger work units had complete sets of living facilities, such as workshops, stores, theatres, restaurants, hotels, hospitals, kindergartens, primary schools, middle schools and even colleges. In effect, they became self-sufficient sub-cities, displaying a typical urban form in socialist China (Yeh and Wu, 1998). In some extreme cases, it was unnecessary for residents to go out of the *danwei* to meet their living demands: complete self-containment. China's socialist self-containment was an effective way to provide for the basic needs of the residents living within a *danwei*, and it worked to alleviate the pressure on limited city-level public facilities in congested metropolitan areas.

In Yangzhou city, to allow land for newly built *danwei*, the municipal government expropriated or demolished previous feudal buildings and historical neighbourhoods. This is also a material consequence of the socialist view of historic places like the royal palace, seen as symbols of the feudal power opposed to socialist principles (Zhu, 1983). A similar situation happened with the wall of Yangzhou city. It was demolished in the 1950s to build a ring road (16–20 m wide) surrounding the old downtown to connect different *danweis*. In the 1960s, to mitigate traffic problems, small rivers connecting previous royal gardens in the downtown areas were filled in to construct wider roads. The Wen river, 18–22 metres in width, is one example (Gao, 1992). These urban constructions not only responded to the new needs of industrialization but also materially transformed the Confucian and imperial urban form to show the dominant position of socialism. The planning practice in that period took account of neither the historic and cultural value of the feudal heritage nor the existing geographical characteristics of Yangzhou city.

These large-scale construction projects in Yangzhou city almost paused during the Cultural Revolution (from 1966 to 1976), when a series of political catastrophes and economic disasters created chaos. Urban-planning practice was therefore halted during this period (Xie and Costa, 1993).

"The road to a user's city" in transitional China

After the Cultural Revolution and the subsequent reform in 1978, the dominance of socialist values and state control on urban planning rapidly diminished. New thinking about social organization and urban lifestyles, housing and land reform, increasing migration and the opening up of Chinese cities to foreign investment resulted in rapid urban development radically different in form from the socialist vision (Shek, 2006). The Chinese central government and management institutions discussed and explored many aspects of the meaning of existing urban ideals, models and templates, including the "garden city" (Ministry of Housing and Urban – Rural Development, 2000), "ecological garden city" (Ministry of Housing and Urban – Rural Development, 2004), and the "livable city" (Ministry of Housing and Urban – Rural Development, 2007). These urban models were based on modern thought and the relevant Western urban theories, such as eco-cities and the rise of the sustainable urban development as a discipline (Zhao, 2011). Each municipal government had its own understanding of those ideal urban models and applied them in a way that factored in the local characteristics and each city's distinctive functions.

Another significant change after the 1978 reform was the announcement of a series of new regulations. In 1984, the Regulation of Urban Planning was adopted by the State Council, which became the guide for urban-planning practice. And in 1989, the Urban Planning Act established legal obligations for the first time in China; a comprehensive master plan was the core of this system (Yeh and Wu, 1998). Prior to this legislation, a master plan was one element of an economic plan, its main function being to assign space for industrial projects. The Urban Planning Act meant master plans became independent spatial plans that could deal with all kinds of land use, including but not limited to industrial developments. Until 2010, the Yangzhou city municipality has made and implemented three master plans: in 1982, in 1996 and in 2002 (Figure 8.1).

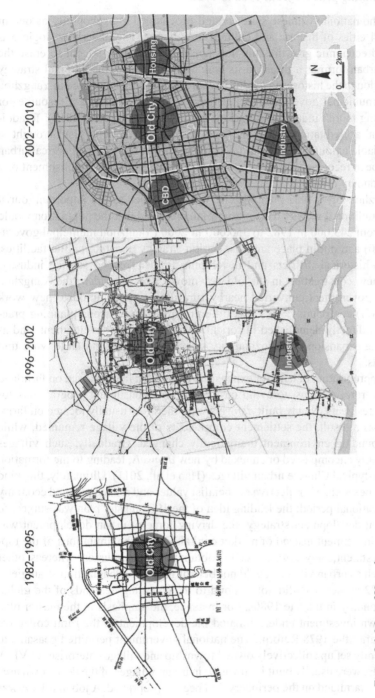

Figure 8.1 Map of Yangzhou

Source: Yangzhou municipal government, 2010, adapted by the authors

Urban-planning practice from 1982 to 1996

In 1982, the national Chinese state council proclaimed Yangzhou city as one of the first 24 cities of historic and cultural importance. The underlying logic was market-led economic growth. This new approach no longer sought to erase the imperial urban form as a threat to its core values. Instead, the national strategy paid attention to the historic and cultural values of the feudal heritage in Yangzhou city. The municipal government recognized cultural heritage as a resource for a developing tourist industry (Yangzhou municipal government, 2010). In addition, formal accreditation as a "state-level historic cultural city" had brought in an appreciable amount of national investment. In this sense, the historical urban form has been recognized as having economic value since the establishment of a market economy in China.

In Yangzhou city's 1982 plan, the city was also a seen as "an important tourist city with traditional views". This master plan predicted that the population would increase from 243,000 in 1982 to 350,000 in 2000 (Yangzhou municipal government, 2010) and put in place an urbanization strategy updating public facilities, protecting historical districts and developing tourism-related service industry. Because new construction in the old downtown was forbidden, the Yangzhou municipal government converted nearby rural land to accommodate new work unit (*danwei*) compounds. This was different from the previous planning practice, which directly demolished historical districts for new development, and as a result, the expansion of Yangzhou city was structured around large work unit compounds.

In the appropriation of land for expansion, the government focused on farmland rather than rural villages, to avoid costly and time-consuming programmes for relocating residents. Newly built *danweis* were therefore usually located on farmland, and as a result, the settlement components of the village remained, while their surrounding environment dramatically changed. Gradually, such villages were spatially encompassed or annexed by new *danweis*, leading to the formation of the now-typical Chinese urban villages (Hao et al., 2011). Ultimately, this kind of spatial coexistence is reflective of socialist values and related planned economy in the transitional period: the leading idea of the 1982 master plan was subject to the national development strategy; the driving force of urban development was top-down investment instead of market capital. Due to the limitations of the top-down investment, only *danweis* could obtain needed resources, whereas other sectors such as urban villages could not benefit from the urbanization strategies.

The 1982 master plan did not respond to the emerging demands of the global market economy in the late 1980s. For instance, the emphasis in the master plan on top-down investment underestimated the development of the rural collective economy after the 1978 Reform. The national government permitted peasants to independently set up collectively owned township and village enterprises (TVEs). These TVEs were usually built in towns or in urban villages of the downtown area because the farmland on the periphery had been expropriated. A job in a TVE was warmly welcomed by peasants, because they could work in local factories and take care of their old rural home if necessary (Li and Rozelle, 2003). Under that model, the urban population in Yangzhou city increased to around 430,000 in the late 1980s, way beyond the estimates of the 1982 master plan. A new master plan that could reflect the global market economy was therefore needed (see Yangzhou municipal government, 2010).

Urban-planning practice from 1996 to 2002

In the 1990s, the 1992 spring tour of Xiaoping Deng (the leader of the party) in the southern provinces created the stage from which to open up and welcome new foreign investment. In 1994 specifically, the tax and fiscal reform encouraged the decentralization of management from higher-level governments to municipal governments. From then on, the national government rarely directly subsidized the development plan of the municipal government (Zhang, 2009). The Yangzhou municipal government was therefore required to attract foreign direct investment in order to respond to the changing macro economy and maintain local economic growth. A common strategy of municipal governments was to establish new riverfront industrial and factory zones where the land rental costs were low or even free. The Yangzhou municipal government employed this strategy while seeking to maintain its status as a state-level historic cultural city.

That dual purpose was reflected in the new master plan in 1996, which sought to protect the old downtown and build a new manufacturing industrial zone near the confluence of the Yangtze River and the Beijing-Hangzhou Grand Canal. After the completion of these big projects, the southern part of the city was well developed, and the urban population increased from 443,000 in 1996 to 556,000 in 2002.

Like most urban industrial zones, the master plan of Yangzhou municipal government dictated the layout of the industrial zone and the provision of public service facilities. Each enterprise leased a piece of land in the industrial zone and built a factory site with connected dormitory housing. To a certain degree, the organization of urban form resembled the way of the *danwei*; indeed, each enterprise became a *danwei* unit in the industrial zone. However, Yangzhou city had acted as a commercial and tourist city for a long time, and consequently, its manufacturing sector was not as competitive as that of other cities, like Wuxi, which had a long history of manufacturing development. Because of this disadvantage, attracting foreign direct investment was not an easy task. As a result, the output of Yangzhou municipality ranked eighth among 13 major municipalities in Jiangsu provinces. The strategy of developing manufacturing industry in the 1996 master plan did not seem to be effective (see Yangzhou municipal government, 2010).

In that light, attracting private investment became an important alternative strategy, and actually, the lack of foreign investment was very common among most cities. To address the problem, the national government promoted the reform of urban housing in 1998 so as to develop a private housing market. It prohibited any new construction of *danwei* housing and sold urban land to developers. People had to purchase private housing (Logan et al., 2009). As a state-level historic cultural city with minimal manufacturing industry and a pleasant environment, Yangzhou city had an advantage over other cities in developing a real estate industry. Realizing this, the Yangzhou municipal government intended to expropriate more farmland via a new master plan and sell it to developers to build market-sector housing. The result was the third master plan, in 2002.

Urban-planning practice from 2002 to 2010

The 2002 master plan predicted that the urban population would reach one million in 2020. Yangzhou city was now officially characterized as a "state-level historic cultural city, and an important tourist destination with traditional scenery and an ecological garden city suitable for living" (Yangzhou municipal government, 2010). The new definition expressed the concerns of the Yangzhou municipal government about the residents' living environment, although it was more or less a compromise with the developing market force: the government needed citizens to invest. To a certain degree, that also reflected a new understanding of the meaning of a city that was much more related to such modern values that suggest cities should be built and organized around the quality of life of their inhabitants.

To improve the living environment, one major strategy of the municipal government was to redevelop historical districts and urban villages and build modernized neighbourhoods in the old downtown or inner suburbs. At the time, thousands of poor migrants lived in private rental housing in downtown urban villages, which resulted in serious social problems such as crime, fire danger, public health hazards and overcrowding (Yangzhou municipal government, 2010). This was the legacy of the 1982 masterplan when no funding had been available to improve the areas surrounding the newly created *danweis* (see 1982–1996 section). The Yangzhou municipal government employed a demolition-redevelopment policy and started the redevelopment of urban villages in 2003 (Yangzhou municipal government, 2003). Old *danweis* and urban villages were replaced with a new tourist district, a new CBD and new residential communities.

Unlike the historic district or urban villages, the new residential communities are gated neighbourhoods with public gardens; buildings are multi-storey instead of one- or two-storey bungalows. This template was very effective. The idea of enhancing the living environment followed the ideal urban models of the "ecological city" and "ecological garden city" encouraged by the national government. Consequently, in 2004 Yangzhou city was presented with the China Habitat Award by the national Ministry of Housing and Urban–Rural Development, and in 2006 the city won the even more prestigious United Nation Habitat Award (Yangzhou municipal government, 2010).

The booming economy prompted hundreds of thousands of rural migrants from less-developed municipalities to arrive in Yangzhou city in the 2000s, so its population increased to 1.2 million by 2010, which was beyond what was anticipated by the 2002 master plan. Without the supply of public housing, the demolition-redevelopment policy of the master plan was unsuited to low-income migrants' needs for housing. As tenants in urban villages, these migrants were neglected by the main interest groups (including the Yangzhou municipal government, developers and landlords – original urban villagers), without reasonable compensation (Hao et al., 2011). They had to move out of demolished urban villages in the downtown or inner suburbs and search for low-rent housing in new urban villages located in outer suburbs. As a result, the settlement pattern of rural migrants shifted from the downtown or inner suburbs to the outer suburbs.

Summary and discussion

This chapter has explored how the transformation of political and economic imperatives shaped the value systems of urban-planning practice in China. The historical account of urban development in Yangzhou city in Jiangsu province highlights a dynamic where a change of the value system altered the meaning of Yangzhou city. In this chapter, we have traced the city's trajectory from a city for the royal family to build a temporary palace in imperial China, through a city as a tool to develop industries in the socialist regime, to the current format in transitional China with the city as an ecological garden and enhanced living environment with an attendant tourist industry.

This change in the meaning of city is reflected each time by the transformation of urban-planning practice. Yangzhou city planning was first based on the Confucian model of an ideal city described in the classical handbook of urban planning, *Kaogongji*, in imperial China. In the socialist regime, the urban development of Yangzhou city was based on the Soviet model – to build socialist neighbourhoods of self-containment, *danwei*; in transitional China, the Yangzhou municipal government applied the modern urban model of an ecological garden city and used the urban renewal practice to improve people's living environment. All these transformations were driven by the dominant power. For a long time, China's urban-planning practice, as a task of government, acted in the interests of those in power to "translate" their values into urban space. The shift of dominant values created distinctive types of spatial form that coexist in urban areas, especially in a historic cultural city such as Yangzhou city. However, China's urban-planning practice, resulting in the limited supply of housing for low-income migrants in transitional China, reveals its neglecting the interests of its most marginalized residents. Although the provision of public facilities and living environment improved after adopting the model for an ecological garden city, it benefited mainly native people with a Yangzhou *hukou* (higher social) status. At the same time, low-income migrants who were rarely able to obtain a Yangzhou *hukou* status hardly enjoyed public amenities such as low-cost public housing and instead had to rely on informal private rental housing or factory dormitories. The demolition of urban villages reduced the supply of private low-cost rental housing in inner-city areas and forced the relocation of low-income migrants to more peripheral areas with less access to public amenities. Hence, we see how urban-planning practice can aggravate social inequality as a side effect. We suggest that the value systems embedded within urban-planning practice and their consequences deserve more attention from urban scholars and policymakers alike.

References

Castells, M. (1983) *The City and the Grassroots: A Cross-Cultural Theory of Urban Social Movements*. Berkeley, Los Angeles: University of California Press.

Gao. D.S. (1992) 'Tan Yangzhou Laochengqu Baohu yu Gaizao Guihua [A study of the urban planning of the old downtown areas in Yangzhou]', *Urban Planning*, 1, pp. 40–43. (in Chinese)

Hao, P., Sliuzas, R. and Geertman, S. (2011) 'The Development and Redevelopment of Urban Villages in Shenzhen', *Habitat International*, 35(2), pp. 214–224.

Ho, P.T. (1954) 'The Salt Merchants of Yang-Chou: A Study of Commercial Capitalism in Eighteenth-Century China', *Harvard Journal of Asiatic Studies*, 17(1/2), pp. 130–168.

Ji, P. (2002) 'Dili Huanjing Bianqian yu Chengshi Jindaihua – Ming Qing yilai Yangzhou Chengshi Xingshuai de Sikao' [Geographic Environment and the Modernization of a City Since Ming and Qing Dynasty: The Case of Yangzhou], *Social Sciences in Nanjing*, 12, pp. 52–55. (in Chinese)

Li, H.B. and Rozelle, S. (2003) 'Privatizing Rural China: Insider Privatization, Innovative Contracts and the Performance of Township Enterprises', *The China Quarterly*, 176, pp. 981–1005.

Logan, J.R., Fang, Y. and Zhang, Z. (2009) 'Access to Housing in Urban China', *International Journal of Urban and Regional Research*, 33(4), pp. 914–935.

Ma, J.C. (2004) 'Economic Reforms, Urban Spatial Restructuring, and Planning in China', *Progress in Planning*, 61(4), pp. 237–260.

Ministry of Housing and Urban – Rural Development. (2000) *Notice on the Issuance of Building "National Garden City" Implementation Opinions*. Available at: www.mohurd. gov.cn/zcfg/jsbwj_0/jsbwjcsjs/200611/t20061101_156922.html (Accessed 27 April 2012). (in Chinese)

Ministry of Housing and Urban – Rural Development, P.R. China. (2004) *Notice on the Issuance of Building "National Ecological Garden City" Implementation Opinions*. Available at: www.mohurd.gov.cn/zcfg/jswj/csjs/200611/t20061101_157113.htm (Accessed 27 April 2012). (in Chinese)

Ministry of Housing and Urban – Rural Development, P.R. China. (2007). *The Standard of Building "Livable City"*. Available at: www.gov.cn/jrzg/2007-06/25/content_660218. htm (Accessed 27 April 2012). (in Chinese)

Shek, D.T. (2006) 'Chinese Family Research Puzzles, Progress, Paradigms, and Policy Implications', *Journal of Family Issues*, 27(3), pp. 275–284.

Smith, M.E. (2007) 'Form and Meaning in the Earliest Cities: A New Approach to Ancient Urban Planning', *Journal of Planning History*, 6(1), pp. 3–47.

Wang, X.Q. (2012) *A Comprehensive Research on Morphological Characteristics of the Architectural Heritage in the Old City of Yangzhou – Taking the Traditional Dwellings as Example*. Unpublished doctoral dissertation, Jiangnan University, China. (in Chinese)

Wu, L.Y. (1986) *A Brief History of Ancient Chinese City Planning*. Kassel: Gesamthochschulbibliothek.

Xie, Y.C. and Costa, F.J. (1993) 'Urban Planning in Socialist China', *Cities*, 10(2), pp. 103–114.

Yangzhou municipal government. (2003) *The Redevelopment Policy of Urban Villages in the Downtown Areas*. Available at: http://wap.yangzhou.gov.cn/mfjxxgk/YZB03/200311/5NOZVP8RQIFQNEHB3WTEL7FIMKTWI67Y.shtml (Accessed 27 April 2012). (in Chinese)

Yangzhou municipal government. (2010) *The Assessment Report of Yangzhou Master Plan*. Available at: http://yzghj.gov.cn/gzcy/ (Accessed 27 April 2012). (in Chinese)

Yangzhou municipal government. (2011) *Yangzhou Statistical Yearbook*. Available at http://tjj.yangzhou.gov.cn/nj2011/INDEX.HTM (Accessed 27 April 2012). (in Chinese)

Yangzhou municipal government. (2012) *The History of Yangzhou City*. Available at http:// daj.yangzhou.gov.cn/daj/yzsz/daj_list.shtml (Accessed 27 April 2012). (in Chinese)

Yeh, A.G.O. and Wu, F. (1998) 'The Transformation of the Urban Planning System in China From a Centrally Planned to Transitional Economy', *Progress in Planning*, 51(3), pp. 167–252.

Zhang, T.W. (2002) 'Challenges Facing Chinese Planners in Transitional China'. *Journal of Planning Education and Research*, 22, pp. 64–76.

Zhang, Y.B. (2009) 'Difang Shenfen Zhixu, Zhufang Huode yu Jingzhengshi Difang Zhengfu' [The identity of locality, home ownership and competition among local governments], *The Journal of Humanities*, 6, pp. 153–160. (in Chinese)

Zhao, J.Z. (2011). *Towards Sustainable Cities in China: Analysis and Assessment of Some Chinese Cities in 2008*. New York: Springer-Verlag.

Zhu, M.W. (1983) 'Yangzhou Guchengqu Dongxi Gandao Gaijian Guihua' [The construction of the traffic infrastructure in the downtown areas in Yangzhou city], *Urban Planning*, 5, pp. 54–58. (in Chinese)

9 Locating value in the Anthropocene
Baselines and the contested nature of invasive plants

Marte Qvenild and Gunhild Setten

Introduction

Degradation, damage and decay all seem to be signatures of the Anthropocene: the geologic period of time in which humans affect all aspects of life on earth seriously. Images of "historical abundance and subsequent decline" (Alagona et al., 2012, p. 49) and a generally "high opinion of pristine wilderness and low opinion of human changes" (Marris, 2011, p. 24) characterize current environmental conservation efforts and policies to such a degree that inherent controversies and contingencies of such efforts often appear to be side-stepped in the quest to save and preserve what might be left of wild, "natural" or undamaged nature. While being commonly portrayed as rooted in disinterested and neutral science, valuation practices and ideologies steering conservation efforts are increasingly exposed and scrutinized (e.g. see Cronon, 1995; Chew, 2009; Bay-Larsen, 2012; Qvenild, 2014). Recognizing the Anthropocene, then, "challenges prevalent and powerful understandings of Nature – a pure and timeless collection of objects, best removed from Society" (Lorimer, 2012, p. 594). In this chapter, we will use the specific example of invasive alien species to locate value as a contested and powerful force shaping nature in the Anthropocene.

Species movements and introductions, particularly those crossing national borders, have contributed to a discursive and material "reshuffling" and, by implication, damaging of Nature. Struggle over (invasive) aliens, or "the bursting out from control of forces that were previously held in restraint by other forces" (Elton, 2000 [1958], p. 15) is born precisely in the Anthropocene. Increased globalized trade and travel have resulted in a massive spread of non-native species, which are seen to severely disturb pristine ecosystems and native species across the world (Low, 2001; Mooney et al., 2005).

As noted extensively elsewhere (e.g. Low, 2001; Alagona et al., 2012; Head, 2012; Qvenild, 2014), the construction of aliens and natives "carries complex interpretive baggage" (Robbins and Moore, 2013, p. 4), hence producing ideologies and practices fraught with challenges. A specific concern in this chapter is to problematize the temporalities and spatialities of historical baselines, or reference states for evaluation, which are inherent to almost all conservation, including the preservation of native species. A baseline separates an alien from a native species and thereby forces perspectives on "before" and "after" ecosystems, which were brought out of balance. Hence, there is an explicit temporality to any conservation effort. Baselines are equally spatial: the movement of species is in itself a spatial activity, which by implication produces conservation measures that often refer to nations or spaces designated for conservation. Finally, and crucially, reference states tend to "become the *good*, the goal, the one correct state" (Marris, 2011, p. 3, emphasis in original). The setting or assumption of a baseline is hence a profoundly value-based and normative activity that has tended to be left unexamined (Bay-Larsen, 2012). A growing body of critique is, however, developing, "indicative of a larger upheaval throughout what could best be described as the Edenic sciences – understood to include, among others, conservation biology, restoration ecology, and invasion biology" (Robbins and Moore, 2013, p. 4). According to Robbins and Moore (ibid.), these sciences "share a tacit epistemological commitment to evaluating ecological relationships explicitly with regard to an a priori baseline". Along with other social scientists, including human geographers, we argue that this only serves to promote an ahistorical "snapshot ecology" that, probably with the best of intentions, idealizes certain past conditions yet fundamentally fails to acknowledge the permeability of the nature–culture divide: "rapid changes in environments around us have [consequently] made the political implications of these sciences harder to ignore or disguise" (Robbins, 2014, p. 104).

Against this background, this chapter offers an exploration of a concrete yet fundamentally value-based and ideological conflict over alien and native species that took place during a massive refurbishment project at the former international airport in Oslo, Norway. Constructed in 1939, the airport was in 1998 relocated from the Fornebu peninsula, and state authorities decided to develop the former landing strips and their surroundings into a site for housing, recreation and business. Soon after the initiation of the project, planners and landscape architects at Fornebu clashed with a conservationist group over which plants to use in the restoration process and to what extent the pre-airport brown field should be restored back to its "original" state. In essence, the conflict was rooted in a heated dispute over baselines for "naturalness", very much represented by the categorization of plants as natives and aliens respectively.

Before moving to the empirical material of Fornebu, we will provide a more-detailed exposition of the increasing acknowledgement of environmental decision-making being fundamentally informed by contested and conflicting value judgements (Gibbs, 2006), including such judgements' differing temporal and spatial effects.

Exploring environmental values

Cresswell (2012, p. 167) aptly argues that "seeing value as intrinsic to any object only hides the interests of those doing the valuing". By implication, the notion of value spans the social, economic, environmental and moral principles shaping our ideas, choices and convictions (ibid.). Value talk and value practice are fundamental to environmental management and policymaking, frequently appealing to our responsibility to demonstrate planetary stewardship. "Environmental value" is, however, a term often used without a clear definition (Gibbs, 2006). Gibbs (2006, p. 74) has distilled two established significations in "environmental value": first, a normative and moral application appealing to justice and responsibility and, second, a utility-based usage closely related to the field of economics. The latter has increasingly made its mark on environmental management and policymaking, such as through the global initiative of The Economics of Ecosystems and Biodiversity (TEEB, 2010). Underlying this approach is a strong normative discourse integral to international environmental policymaking stressing our responsibility to protect and conserve the environment for its own sake. The former appeals to our moral obligations to save an environment in peril – one of the signatures of the Anthropocene. Seeing the environment as in need of help – that is, to again become healthy, safe, pure or more authentic (Lowenthal, 2010) – often means to protect or restore in isolation from the processes that produced it. In effect, carefully selected spaces or species for conservation are claimed to have intrinsic value independent of their use value (Castree and Braun, 2001). At the core of this rhetoric lies the idea that environmental values rest on objective and neutral scientific assessments (Robbins and Moore, 2013). This is a de-contextualized and atemporal perspective, and in line with, for example, Cresswell (2012), we hold that estimating a value is never intrinsic and pre-given but is rather an emerging process. Such a perspective holds that values are produced, reproduced and negotiated between those doing the valuing with a reference to what is being valued. The practices and politics of discriminating between alien and native species, including setting baselines, demonstrate the negotiating of values ascribed to the environment.

Baselines in restoration ecology

The setting of reference states or baselines is a common tool in restoration ecology for defining natural conditions where there is an "absence of significant human disturbance or alteration" (Stoddard et al., 2006, p. 1267). On the one hand, baselines are useful in assessing environmental change, for example, and can be used by policymakers to evaluate the effectiveness of environmental policies and management (Roe, 2013). On the other hand, however, and crucially, there are no *correct* baselines. Daniel Pauley (1995) pointed out how the setting of "wrong" baselines led to a degradation of marine fish stocks, through what he termed the "shifting baseline syndrome" (Pauley, 1995, p. 430). This "syndrome" referred to how succeeding generations of fisheries scientists used the fish-stock size found at the beginning of their careers as a baseline for evaluating change. The broader relevance of Pauley's (1995) work is that inherent to shifting – that is, incorrect baselines – is a "lowering of standards of nature and the acceptance of degraded natural ecosystems to be the normal state of nature" (Vera, 2010, p. 98). As a result, then, by demonstrating how fish stocks continued to decline towards the rate of extinction, Pauley (1995) made a general call for increasing the use of historical records to set "true" or correct baselines.

If we follow Pauley (1995, see also Marris, 2011), the truth of nature lies in the past: in a "natural state" before contaminations or impacts occurred (e.g. Willis et al., 2003; Stoddard et al., 2006; Vera, 2010). What we find problematic is that the scientific setting of baselines often appears to be undertaken without an examination of the values that these reference states rest on: "Baselines do not usually contain an evaluation of the survey data; value judgements are not made", as Roe (2013, p. 771) has bluntly stated.

There is, however, a growing body of literature broadly related to ecology that addresses the limitations and uncertainties associated with the setting of baselines for restoration (e.g. see Willis et al., 2003; Burger et al., 2007). Crucially, as ecosystems change and have always been in flux, historical data will always be fragmentary and incomplete (Alagona et al., 2012). Duarte et al. (2009) argue that rather than searching for idealized nature states, scientists, conservation managers, policymakers and interest groups need to acknowledge shifting baselines and facilitate sustainable ecosystems that can handle change rather than try to recreate static idealizations of a past. Despite such much needed critical reflections over the limitations of baselines, little insights have spilled over into the field of species conservation and politics where alien, human-introduced species are distinguished from native and original species in often rather arbitrary ways. We therefore need to take more seriously the effects of what Low (2001, p. 36) holds when he points out that "our values contribute to our pest problems".

Baselines producing alien and native species

Baselines set in order to distinguish between alien and native species are commonly seen in policymaking as a clear-cut spatial and temporal separation between which species belong and which do not (Qvenild, 2014). On a closer look, baselines appear to be set rather arbitrarily (cf. e.g. Marris, 2011; Head, 2012). First, they are set because it often proves difficult to determine the natural range of different species, including their mode of introduction: did the species arrive at a location by its own means or have humans introduced it at some point? Second, they are set because a great variety of temporal baselines between nativeness and alienness attests to the aforementioned "complex interpretive baggage" (Robbins and Moore, 2013, p. 4) of current species debates. In Australia, the year 1788 has been set as baseline because it marks the start of the European colonization (Head, 2012): a fundamentally ideological boundary is drawn. Examples of other baselines are the Neolithic period; the last Ice Age, ten thousand years ago; and the closure of the English Channel, seven thousand years ago.

In Norway, the year 1800 has become a year zero, meaning that species introduced to Norway after 1800 are categorized as aliens. The year 1800 consequently marks which species movements and human practices are deemed acceptable or not. An alternative year of 1750 has been set for vascular plants because it marks the beginning of a more large-scale introduction of alien plants to Norway. The years 1800 and 1750 respectively are, however, chosen primarily for regulatory purposes and for providing the authorities with a possibility to distinguish what is *assumed* to belong in Norwegian nature from that which is not. The value-based activity of setting these baselines was largely left unexamined by the scientists who contributed in the process of compiling what became the first Norwegian Black List of unwanted species published in 2007, with an updated list published in 2012 (Qvenild, 2014). The intention with black listing and the more commonly red listing species is to inform decisions over which species to conserve or protect and which to eradicate. Currently, Norwegian environmental authorities are working on a new legal regulation to prohibit the sale of black-listed alien plants. Such a systemized and hierarchized representation of nature forces a critical discussion of practices resulting from the formal establishment of degraded nature. And as Marris (2011, pp. 11–12) aptly states, "To hold the blocks to a simulacrum of 1770, conservationists must shoot, poison, trap, fence, and watch, forever watch, lest the excluded species find their way back in. [Hence] A historically faithful ecosystem is necessarily a heavily managed ecosystem" and is not quite what many conservationists hold as an ideal. In this chapter, we have set out to critically examine how values of species are produced and reproduced, and it is consequently due time to turn our focus towards Fornebu in Oslo and look more closely at how different species were mobilized in clashing efforts to (re-)establish a vision of a natural site.

A brief note on methodology

Qualitative data were collected through 18 semi-structured interviews with key people involved in the restoration process at Fornebu. The interviewees have been categorized into two groups, representing the conflicting parties; landscape architects and planners employed by and associated with Statsbygg (The Norwegian Directorate of Public Construction and Development) and conservationists associated with SABIMA (The Norwegian Biodiversity Network).[1] These are rough categories, but they do illustrate the point we want to make. The majority of interviews were undertaken as talking-while-walking interviews (cf. e.g. Hitchings and Jones, 2004), which enable an active reflection on and engagement with the contested plants and the wider landscape. In addition, a content analysis of relevant documents was undertaken, including such documents as statutory plans, background documents, reports, leaflets, personal letters, email correspondence and media coverage. Inspired by Skinner (2002), the analysis focused on how the parties defined and applied key concepts such as alien, native and original – that is, how concepts were set to do work in conflict situations.

Fornebu airport – 70 years of disturbance?

> The reestablishment of original nature and conservation of fragile areas did not take place at all [at Fornebu]. . . . The restoration was so poorly undertaken that it was reported to the police.
>
> (Knudsen, 2012, n.p.)

Following the relocation of the airport in 1998, Statsbygg, acting on behalf of the Norwegian state, took responsibility for a massive redevelopment project at the former brown fields. Sustainable Fornebu was launched and was planned to host over six thousand dwellings and 20,000 offices in addition to a refurbishment of areas for biodiversity, wildlife and recreation. Refurbishing Fornebu was in Statsbygg's own terminology one of Norway's largest and most ambitious development projects, where the underlying aim was to become a showcase for modern environmental thinking by emphasizing sustainability in all phases of the process (Statsbygg, 2002). Leaning on a vision of a pre-airport landscape, which was characterized by a mosaic of different types of forests, swamps and flower meadows with diverse flora and fauna, including over 260 bird species (Bendiksen, 1994), Statsbygg was designated a mandate to conserve and restore both cultural and natural elements of this landscape. Included in the mandate was a requirement to use native plants in the restoration process (Østengen and Bergo, 2001).

Testimony to the natural values of the area is that the wetlands were protected as nature reserves by royal proclamation in 1992 because of their importance for migrating birds. Local conservationists, who had been using the area for bird watching during the airport period, were lobbying in the planning process to make sure that the rich bird life and the wider qualities of the area were properly addressed. By the end of the project, the conservationists realized, however, that the landscape had turned into a "sterile park landscape" and not the dry meadows and calcareous pine forests that they had imagined (Bergan, 2009). As a result, the conservationists reported Statsbygg to the police in 2007 for having violated the regulatory plan through the plantings of alien rather than native species (Bærum kommunestyre, 2002). The police closed the case quickly due to a lack of investigating capacity. What interests us, though, is the fact that the conservationists had been quite satisfied with the initial planning documents for Sustainable Fornebu but were largely disappointed by the final results (Qvenild, 2014). What happened, then, in the realization of the statutory plans?

Protecting biodiversity in general and the bird life in particular in the two nature reserves were key principles in the project. Consequently, the planting of fast-growing plants to form dense buffer zones in order to protect the bird life from the new residents and office workers was highly prioritized (Bjørbekk and Lindheim, 2005). Vegetation was also key to the establishment of the green corridors that were to connect the two reserves and ensure that wildlife could thrive. Difficulties began, however, with the rather vague definition in the central planning documents of *native* species – that is, vegetation growing in the inner Oslo fjord (Bærum kommune Rådmannen, 2001; Bærum kommunestyre, 2002). The lack of a specified baseline for belonging led the planners and landscape architects to believe that species like the Japanese rose was acceptable to plant at Fornebu because it had been growing on neighbouring islands in the Oslo fjord for decades (Bjørbekk and Lindheim, 2005). The planners hence made a working definition of native as "species which today grow at Fornebu and which includes both original, natural species diversity and cultural plants which have become naturalized or are part of the cultural landscape" (Lundetræ, 2007, n.p.). Moreover, they treated the statutory requirement of planting native species with pragmatism. The fast establishment of the green buffer zones was their main priority, and when the appropriate native species were hard to produce or get hold of on the commercial market, these were substituted by similar non-native commercial species (Statsbygg and Oslo Kommune, 2000).

The conservationists were outraged over the result. Their definition of native species was species that had been growing at Fornebu before 1750 or also 1820, when a number of alien plants were introduced to Norway through the modernization of agriculture (Qvenild, 2014). Further, the landscape architects and the conservationists disagreed over the invasive potential of the alien species. While the conservationists feared the spread of the newly planted alien species to the nature reserves, the planners and landscape architects argued that species like Japanese roses were not invasive on neighbouring islands, and moreover, these plants were also found in nearby gardens – that is they were native and did not spread.

Not only did the conservationists have strong reactions to the planting of invasive alien species. They were also disappointed by how the landscape had been (re-)created. In the statutory plans, a pre-airport landscape (i.e. before 1939) served as a baseline for the development of the new landscape. Which version of a past landscape that was going to be re-created was far from clear. The planners and landscape architects studied old maps and were inspired by the former agricultural landscape and were hence able to justify their use of nutrient-rich soils. In fact, Statsbygg had never intended Fornebu to be a restoration project, because the area had been heavily contaminated and polluted due to the construction and running of the airport. Statsbygg's aim was to create a new, clearly human-made landscape and to conserve the remaining patches of vegetation and parts of the former landing strip.

In sharp contrast, the conservationists envisioned other parts of the pre-airport landscape to be both conserved and restored. They imagined a mosaic landscape of pine forests and dry meadows where drought-tolerant red-listed species could thrive. Moreover, they saw Fornebu as an opportunity to reverse the trend of habitat destruction in the increasingly urbanized Oslo fjord region. In 2007, when the conflict at Fornebu culminated in the police report, the Norwegian Black List for alien species was released, and several of the species planted at Fornebu, including the Japanese rose, were in fact listed as invasive and ecologically harmful species. The conservationists used the black list to strengthen their casting of Statsbygg as an environmental criminal. Importantly, they got extensive coverage in local and national media (see e.g. NRK Lørdagsrevyen, 2007), and as a result, Statsbygg was forced to remove several of the black-listed species and by implication were publicly judged as behaving environmentally irresponsibly.

Given this story, it is rather surprising that Statsbygg's dubious practices seem to have been almost forgotten seven years later. A quick Google search with the keyword "Fornebu" gives the impression of an attractive place to live and visit for recreational purposes – which it might very well be. Statsbygg's Sustainable Fornebu was, in fact, one of the winners of the prestigious European and Regional

Planning Awards in 2014. The jury of the award were particularly impressed by how the Fornebu project combined sustainable energy solutions with "a broadly based strategy for safeguarding and strengthening the biological diversity and landscape qualities of the area". Those with an attentive eye will, however, observe a number of invasive alien plants spreading and establishing at Fornebu, including the Japanese rose. The grounds on which the award was based do not mention the extensive use of invasive alien plants. Interestingly, this unfolds at a time when rules and regulations against the spread of such species are becoming stricter in most countries, including Norway. One explanation of why Fornebu has largely escaped being labelled an ecological catastrophe is most probably due to the rather arbitrary baselines set to distinguish alien from native species, because nature is simply a moving target and increasingly ecologically novel. In most cases, it will be impossible for those untrained in ecology or biology to judge whether a species *belongs* or not (e.g. Qvenild, 2014). More often, and popularly, plants are cherished garden ornamentals that are appreciated for their beautiful flowers (Qvenild et al., 2014). Another explanation may be that traditional ecological ethics and tools, such as the setting of baselines, "confirmed" through the current practice of listing species, may not be well suited to sort out matters in the Anthropocene (e.g. Robbins, 2014). In the final section, we return to these more principal issues.

Concluding remarks

This chapter has examined how the past informs the future through the case of baselines set for distinguishing alien from native species. Such temporal baselines have spatial consequences, including raising complex questions of values and culpability. We argue that the setting of baselines is often done without examining the values attached to such reference points. Baselines in environmental policy-making and in landscape planning may appear clear-cut on paper, but their fuzzy and slippery nature complicates practical and real-world projects where differing expectations and visions clash. As we have demonstrated, Fornebu serves as a striking example of how conflicting values are shaping environments, spatially, temporally and symbolically.

The larger picture concerns a current dilemma: "We want a world in which people are as free as possible to travel and to exchange goods and ideas. . . . But at the same time, we *need* a world in which most other living things stay put" (Low, 2001, p. 41). Not everyone would agree on this, of course, yet Low (ibid.) points at the critical works of fundamental values that can be seen to dominate the Anthropocene, not to say producing the Anthropocene itself. Those values that contribute "to our pest problems include love of mobility, freedom, speed, diversity, progress, familiarity, and a mechanistic view of nature" (ibid., p. 37). All these come together at Fornebu, demonstrating the growing challenge and possibility of returning to native or natural states of environments in the Anthropocene. Most native plants are in fact associated with and dependent on cultivation. This was also the case at Fornebu, where the dry meadow plants cherished by the conservationists were associated with the former cultural landscape and in need of mowing for their survival. However, many landscapes, like that of Fornebu, have been so severely altered that restoring them back to past conditions is impossible (Robbins, 2014). Believing in going back is hence, according to Duarte et al. (2009), not much unlike believing in the existence of Peter Pan's *Neverland*. Robbins thus urges us to reflect on the following: "If the landscape is not going 'back', and the role of the restorations is increasingly revealed to be one of *design* as much as recovery, then upon what basis are we to make choices about future environments?" (2014, p. 111, emphasis in original). Fornebu and the case of alien and native species illustrate the importance of the political works of value choices in a planning project and illustrate that maintaining, restoring or producing "natural" diversity in the Anthropocene lend itself to active intervention. Because the past is ambiguous and unstable, yet portrayed and used as if undisputed, future environments will by default be the results of whatever choices we are able to make, where and with reference to which point in time.

Note

1 SABIMA is an umbrella NGO working to strengthen the protection of biodiversity in Norway. Around 20,000 people hold memberships, covering both professionals and amateur biologists in Norway.

References

Alagona, P.S., Sandlos, J. and Wiersma, Y.F. (2012) 'Past Imperfect: Using Historical Ecology and Baseline Data for Conservation and Restoration Projects in North America', *Environmental Philosophy*, 9(1), pp. 49–70.

Bærum kommune Rådmannen. (2001) *Estetiske retningslinjer for Fornebu* [Aesthetic guidelines for Fornebu]. Bærum: Bærum kommune.

Bærum kommunestyre. (2002) *Reguleringsplan for Storøya* [Regulatory plan for Storøya]. Bærum: Bærum kommune.

Bay-Larsen, I. (2012) 'The Premises and Promises of Trolls in Norwegian Biodiversity Preservation. On the Boundaries Between Bureaucracy and Science', *Environmental Management*, 49(5), pp. 942–953.

Bendiksen, E. (1994) *Botaniske undersøkelser på Fornebu. Vurdering av naturområder i forbindelse med endret arealbruk* [Botanical surveys at Fornebu. Evaluation of nature areas in relation to changes in land use]. Oslo: Norwegian Institute for Nature Research.

Bergan, M. (2009) Hva gikk galt på Fornebu? [What went wrong at Fornebu?]. *Nøttekråka Membership Magazine for Friends of the Earth Bærum*, 1, pp. 3–9.

Bjørbekk, J. and Lindheim, T. (2005) *Forprosjekt for Sentralparken Fornebu Oslo* [Pre-project for the Central Park at Fornebu, Oslo]. Oslo: Bjørbekk & Lindheim.

Burger, J., Gochfeld, M., Powers, C.W. and Greenberg, M. (2007) 'Defining an Ecological Baseline for Restoration and Natural Resource Damage Assessment of Contaminated Sites: The Case of the Department of Energy', *Journal of Environmental Planning and Management*, 50(4), pp. 553–566.

Castree, N. and Braun, B. (2001) *Social Nature. Theory, Practice and Politics*. Oxford, Massachusetts: Blackwell.

Chew, M. (2009) 'The Monstering of Tamarisk: How Scientists Made a Plant Into a Problem', *Journal of the History of Biology*, 42(2), pp. 231–266.

Cresswell, T. (2012) 'Value, Gleaning and the Archive at Maxwell Street, Chicago', *Transactions of the Institute of British Geographers*, 37(1), pp. 164–176.

Cronon, W. (1995) 'The Trouble with Wilderness; or, Getting Back to the Wrong Nature', in Cronon, W. (ed) *Uncommon Ground. Toward Reinventing Nature*. New York, London: W.W. Norton & Company, pp. 69–90.

Duarte, C.M., Conley, D.J., Carstensen, J. and Sánchez-Camacho, M. (2009) 'Return to Neverland: Shifting Baselines Affect Eutrophication Restoration Targets', *Estuaries and Coasts*, 32, pp. 29–36.

Elton, C.S. (2000) [1958]. *The Ecology of Invasions by Animals and Plants*. Chicago: University of Chicago Press.

Gibbs, L.M. (2006) 'Valuing Water: Variability and the Lake Eyre Basin, Central Australia', *Australian Geographer*, 37(1), pp. 73–85.

Head, L. (2012) 'Decentring 1788: Beyond Biotic Nativeness', *Geographical Research*, 50(2), pp. 166–178.

Hitchings, R. and Jones, V. (2004) 'Living With Plants and the Exploration of Botanical Encounter Within Human Geographic Research Practice', *Ethics, Place & Environment*, 7(1–2), pp. 3–18.

IUCN Council. (2000) *Guidelines for the Prevention of Biodiversity Loss Caused by Invasive Alien Species*. Gland: International Union for Conservation of Nature.

Knudsen, K. (2012) *Hva er erfaringene med Fornebu sett i tilbakeblikk?* [What are the retrospective experiences from Fornebu]. Available at: http://lag-ab.nofoa.no/Erfaringene-med-Fornebu.pdf (Accessed 6 May 2014).

Lorimer, J. (2012) 'Multinatural Geographies for the Anthropocene', *Progress in Human Geography*, 36(5), pp. 593–612.

Low, T. (2001) 'From Ecology to Politics: The Human Side of Alien Invasions', in McNeely, J.A. (ed) *The Great Reshuffling. Human Dimensions of Invasive Alien Species*. Gland, Switzerland and Cambridge: IUCN, pp. 35–42.

Lowenthal, D. (2010) 'Reflections on Humpty-Dumpty Ecology', in Hall, M. (ed) *Restoration and History. The Search for a Usable Environmental Past*. New York: Routledge, pp. 13–34.

Lundetræ, V. (2007) '*Definisjon av begrepet 'stedegen vegetasjon*' [*Definition of the term 'native vegetation'*]. Oslo: Memorandum. Statsbygg.

Marris, E. (2011) *Rambunctious Garden. Saving Nature in a Post-Wild World*. New York: Bloomsbury.

Mooney, H., Cropper, A. and Reid, W. (2005) 'Confronting the human dilemma', *Nature*, 434(7033), pp. 561–562.

NRK Lørdagsrevyen[Norwegian Broadcasting Corporation]. (2007) Feil planter på Fornebu [Wrong plants at Fornebu] Television Broadcast. May 27. Oslo: Norwegian Broadcasting Corporation.

Østengen, K. and Bergo. J. (2001) *Landskapsplanen for Fornebu* [*Landscape Plan for Fornebu*]. Oslo: Statsbygg, Oslo kommune og Bærum kommune.

Pauley, D. (1995) 'Anecdotes and the Shifting Baseline Syndrome of Fisheries', *Trends in Ecology and Evolution*, 10(10), p. 430.

Qvenild, M. (2014) 'Wanted and Unwanted Nature: Landscape Development at Fornebu, Norway', *Journal of Environmental Policy & Planning*, 16(2), pp. 183–200.

Qvenild, M., Setten, G. and Skår, M. (2014) 'Politicising Plants: Dwelling and Invasive Alien Species in Domestic Gardens in Norway', *Norsk Geografisk Tidsskrift-Norwegian Journal of Geography*, 68(1), pp. 22–33.

Robbins, P. (2014) 'No Going Back: The Political Ethics of Ecological Novelty', in Oka-moto, K. and Ishikawa, Y. (eds) *Traditional Wisdom and Modern Knowledge for the Earth's Future*. International Perspectives in Geography, Tokyo: Springer, pp. 103–118.

Robbins, P. and Moore, S.A. (2013) 'Ecological Anxiety Disorder: Diagnosing the Politics of the Anthropocene', *Cultural Geographies*, 20(1), pp. 3–19.

Roe, M. (2013) 'Policy Change and ELC Implementation: Establishment of a Baseline for Understanding the Impact on UK National Policy of the European Landscape Conven-tion', *Landscape Research*, 38(6), pp. 768–798.

Skinner, Q. (2002) *Visions of Politics: Regarding Method* (Vol. 1). Cambridge: Cambridge University Press.

Statsbygg. (2002) *From Airport to Sustainable Community: Sustainable Fornebu*. Avail-able at: www.statsbygg.no/Utviklingsprosjekter/Fornebu/Publikasjoner–Fornebu/ (Accessed 5 January 2010).

Statsbygg & Oslo kommune. (2000) *Etterbruk Fornebu Grøntstrukturplan* [Re-development of Fornebu Green structure plan]. Oslo: Statsbygg og Oslo kommune.

Stoddard, J.L., Larsen, D.P., Hawkins, C.P., Johnson, R.K. and Norris, R.H. (2006) 'Setting Expectations for the Ecological Condition of Streams: The Concept of Reference Condi-tion', *Ecological Applications*, 16(4), pp. 1267–1276.

TEEB. (2010) *The Economics of Ecosystems and Biodiversity. Mainstreaming the Eco-nomics of Nature. A Synthesis of the Approach, Conclusions and Recommendations of TEEB*. Malta: Progress Press.

Vera, P. (2010) 'The Shifting Baseline Syndrome in Restoration Ecology', in Hall, M. (ed) *Restoration and History: The Search for a Usable Environmental Past*. New York: Routledge, pp. 98–110.

Willis, R.D., Hull, R.N. and Marshall, L.J. (2003) 'Considerations Regarding the Use of Reference Area and Baseline Information in Ecological Risk Assessments', *Human and Ecological Risk Assessment: An International Journal*, 9(7), pp. 1645–1653.

10 "And what do you do with five-hundred million stars?"

Assessment of darkness and the starry sky, values and integration in regional planning

Samuel Challéat and Thomas Poméon

Introduction

"How can we own the stars?" asks Little Prince the businessman in Antoine de Saint-Exupery's famous 1943 French novella. Behind this innocent interrogation is a less innocent question: can we own the stars? If this is not possible literally, with the possible exception of the closest celestial objects like the moon,[1] objects (here the starry sky) are nonetheless appropriated through use and the enjoyment and well-being it generates – a use that can be associated with monetary and non-monetary values. The prince's question might also implicate a suite of economic transactions, investments and costs involved in the preservation of the object and, by association, a range of techniques and strategies to manage the access and use of related goods and services. Here we enter the field of the economy, with its property rights, accumulation regimes, values, prices and other regulation devices brought in to manage the access and use of the valued object.

This chapter explores relations between light and non-light in its configuration as the night sky as an object increasingly incorporated into systems of valuation through planning policy, preservation initiatives and tourist-linked nocturnal activities. There has been a great deal of work across the social sciences on the damaging effects of darkness in terms of crime and urban vulnerability (Atkins et al., 1991) and the way the night is culturally associated with negative meaning in ethics and aesthetics (the many symbolic oppositions between light and darkness in the Abrahamic religions are examples, among others). Yet, there has been little research on the social, cultural and economic value of the night and its associated qualities of darkness. However, recent scholarship that considers the night sky (NS) as a common but endangered resource and potentially part of a place's tourism promotion leads us to wonder about its value at large. Several authors have recently tackled this question. Gallaway (2010) argued that night sky was undoubtedly an important cultural asset but one needing the recognition of the instrumental aesthetic value of natural beauty. Other works focused on the monetary value of the night sky, using a range of economic methods and concepts, including revealed preference, contingent valuation and willingness to pay (Simpson and Hanna, 2010; Besecke and Hänsch, 2015; Willis, 2015). Despite the methodological and conceptual limits (the many critiques of contingent valuation, for instance, or the difficulty of pinpointing exactly what it is that people valued about a dark and starry night sky), these works have helped us understand what is at stake and have brought attention to the night sky as a resource.

Work remains to be done to analyse processes of night sky valuation at a collective level and to consider the geographical variation in the way the dark and starry sky is recognised, protected and valorised. We present preliminary findings from a broader project about the night sky conducted within the RENOIR research group on night, tourism and rural dynamics. The research includes different field experiences: participant observation during light festivals (Lumiville congress and Fête des Lumières in Lyon, France) and events around the nocturnal environmental resources (star parties and astronomy events, Nuit des Étoiles), comprehensive interviews (International Dark-Sky Reserves of the Pic du Midi de Bigorre, France, and of the Mont Mégantic, Québec) and a content analysis of the justification schemes (see Dupuy et al., 2015).

We introduce the idea of the night sky as a resource by identifying some of the benefits attributed to darkness and, conversely, problems associated with the lack of darkness: *light pollution*. We explore how the night sky is being appreciated and integrated into contemporary urban and rural planning policy at various scales and examine some of the tensions in how the night sky gets incorporated into projects of valuation, such as cost-benefit analyses or ecosystem service frameworks.

The recent consideration of the night and the night sky cannot be separated from the proliferation of artificial lighting (Riegel, 1973; Narisada and Schreuder, 2004; Challéat, 2010; Edensor, 2015). Indeed, values assigned to the night and the starry sky became apparent only through their absence, caused by the excessive use of artificial light in the urban industrial era. This is how we came to talk first about nuisance and subsequently light pollution. Light pollution refers to undesired artificial, mostly electric lighting effects (including such things as glare, sky glow and light trespass). In other words, it is through the loss of the dark night and the discovery of the disadvantages of artificial lighting that emerge the "positivities" now attributed to the dark night and the starry sky to which it provides access. This concern to protect an environmental object following its degradation is quite familiar. It is found, for example, in the emergence of concerns about ecosystem services and biodiversity.

Such reflection enables us to (re-)question the extent and limits of environmental economics since the object of concern, the night and the starry sky, are a dynamic complex combination of nature and society. They have a quasi-immateriality uncommon to much of the more discrete and locatable components in the environment subjected to valuation projects by conservation sciences, and they prompt us to reconsider notions such as ownership, uses, services, externalities and the values they are granted. Gallaway et al. (2010) connected artificial light intensity (and thus light pollution) to economic factors, showing that GDP and population correlate closely with this kind of pollution. In this chapter, we discuss how the absence (or low level) of artificial light from streets, factories, commercial and residential sources can be turned into a valuable asset for less-developed rural areas.

In the first section, we situate the night and the night sky in its prevailing economic logics of environmental values. We then move on to show how the night is appropriated for the production of services in and of the night sky (and hence associated benefits and values are put to use). We focus finally on a specific project of valuation of the night sky, with the creation of dark-sky protection areas and its implications for natural, cultural and economic relations of the region.

What values are accessible through the night?

The first question we must ask is, how is the night sky positioned within an economic typology of goods? As an open access and non-rival resource, the night sky may be considered as a pure public good, following the classification of Ostrom and Ostrom (1977). Such pure public goods are inherently not amenable to market-driven mechanisms of management and regulation, and their economic value is difficult to assess. Indeed, the night sky is not a commodity (like petroleum or wheat) with a price reflecting the dynamic between supply and demand. One way to approach its value is to consider the use and the benefits we derive from it as a shared common resource. This approach falls within the instrumental value paradigm (Brown, 1984; Chan et al., 2012). As shown by Gallaway (2010) and Challéat (2010), all biological life, and humankind in particular, has a long-running and profound connection with the starry night sky. It has been used as a tool and a resource for scientific, artistic or recreational purposes. Stellar navigation, the measurement of time and seasons, inspiration for painting, stargazing and astronomy are all examples of activities based on the night sky. For a long time, the question of its value was not at stake, because its access was easy and independent of human management and action, and thus priceless. This situation has changed during the last two hundred years. Modern urban industrialisation and the incremental loss of the night sky – its scope and intensity – has led to a recognition of its values (Challéat and Lapostolle, 2015).

Public goods, even if inclusive and non-rival, can be subject to dispute and conflict either through their use, misuse and exploitation or because of deleterious effects by an external activity (pollution). In the case of night sky, these conflicts stem from the fact that the night, which allows access to night sky and its related resources/benefits, is also used as a frame good: necessary for lighting activities and business (with objectives that range from security lighting to place enhancement). Certainly, street lighting is also a pure public good in terms of visibility, safety and the extension of human activity. However, for night sky advocates and increasingly for policymakers, street lighting can also be conceived as a negative externality, described and objectified scientifically as nuisance or pollution. In this new frame of reference, street lighting can become an object of cost saving and energy-use reduction, which needs to be balanced with safety and security. The notion of "getting lighting right" (to light just what is needed, where it is needed, when it is necessary), which made its appearance in the doctrine of public lighting in the early 2010s, reflects this evolution. It opens the way to new planning compromises that redefine the balance between the human and nonhuman needs of light and darkness in the nighttime city. This reflects the shift from a functionalist conception – that is, the segmentation of space by urban-planning experts into zones dedicated to housing, employment, traffic, leisure and consumption – towards a sociotechnical approach to urban lighting, which adapts lighting techniques to the multiplicity of uses in the city (Challéat and Lapostolle, 2018).

Access to night sky is linked to the quality of the natural darkness and therefore to light pollution. So the issue is primarily a problem of preservation and of managing negative *externalities* (as economists say). In this sense, a framework related to the conservation and enhancement of natural resources and environmental goods, such as the framework of ecosystem services (MEA, 2005), seems to be the favoured option for current valuations of the night sky within economics. While the ecosystem services framework has been criticised for its limited instrumental and monetised focus and the sub-consequent reduction of nature it implies (Ernstson and Sörlin, 2013; Kull et al., 2015), it can help to describe and analyse the complex relation between night sky, as an object of nature, and humanity. Moreover, as a present dominant paradigm, it influences the way the issue of night sky protection and valuation is tackled by different actors, especially in the political field.

The Millennium Ecosystems Assessment (MEA) framework led to a greater recognition of the need to manage the environment in a more sustainable manner. From this instrumental point of view, the question becomes, which services, benefits and values does humankind derive from the night sky (and more broadly from the night)? And, consequently, which services, benefits and values are lost with light pollution?

The darkness of the night can be seen as a supporting service (or a function), because it is at the heart of fundamental natural cycles necessary for the functioning of the biosphere (alternating day/night), which affects other types of ecosystem services (provisioning, regulating and cultural services) (MEA, 2005). More specifically, several authors (Narisada and Schreuder, 2004; Marín and Jafari, 2007; Challéat, 2010; Gallaway, 2010, 2015) have highlighted a range of services and benefits related to the night sky and darkness and those impinged on by light pollution. According to MEA (2005) and Chan et al. (2012), the classification the night sky and darkness provides many cultural services from different categories (outdoor recreation services, educational and research nature-based services, artistic nature-based services, ceremonial place-based services) and benefits (aesthetics, inspiration, spirituality, knowledge, employment, identity, social capital and cohesion). Natural darkness also provides regulating services related to (wild) fauna and flora (for predation, reproduction, migration) and to human health (hormonal regulation by melatonin, with its chronobiotic function, is a major time setter).

These services are not produced by human devices: the benefits they generate are a priori available everywhere during nighttime. But this a priori availability is compromised by artificial lighting, which produces both benefits and drawbacks.

We cannot deal with this issue by computing costs and benefits without considering the question of scale. Indeed, if the provision of the night sky is equally distributed across the globe (with variation depending on geographical situation and seasonal cycle), received benefits are more local. Similarly, the positive or negative effects of artificial lighting on human well-being implies different scale dimensions and spatial trade-offs. While the benefits of artificial lighting are generally local (increased visibility, sense of safety, showcasing a monument), with little distance between the site of illumination and the targeted beneficiaries, darkness benefits are more disparate, depending on the impact of different kinds of light pollution. Light pollution introduces spatial disparities; it erodes the night, its duration and intensity, differently in different areas (urban vs rural areas, economically developed vs less-developed areas, etc.). If a misguided street lamp directly affects sleep for the neighbourhood and the weasel, the sky glow produced by a city affects the territory of tens or even hundreds of miles around. Thus, cumulatively, the street lamp can affect migratory birds, night sky visibility and the nightscape. We must also acknowledge the indirect effects of artificial lighting: the energy required and the associated global environmental impacts it has.

Such spatial disparities present a real challenge for public policy formulation. They imply different levels of decision (local, regional, national and international), as a local agreement to diminish light pollution can be affected by an extra-local source of artificial lighting (e.g. starry sky visibility of a mountain area can be affected by the sky glow of a 100-mile distant metropolitan area). In France, after conflicts and mobilisations (led by astronomers first and joined later by ecologists) (Challéat and Lapostolle, 2015), the public authority is now aiming, as part of its environmental policy, to reduce the overall impact of light pollution. This policy change is linked to several dynamics, pushing to reduce light pollution: the direct environmental impact of excess artificial light, carbon emissions, energy costs and changes in technology that enable new lighting methods and opportunities for lighting businesses (e.g. LEDs and automated sensor lighting). Other developed countries are also working in similar ways at the national level. At a more local level, a specific protection policy for some areas is emerging with restrictions on the kinds of lights used, their location and their operating hours. This approach works to acknowledge and enhance the night sky as a new resource (a comparative advantage) for regions seeking to benefit from their marginal rural location.

Night sky–friendly regions meet with the need to reach an acceptable trade-off between the benefits of the artificial light and those provided by night sky. This arbitration can be fed by a cost-benefit analysis, which involves the qualification and quantification of values associated with each service and benefit. The values associated with artificial lighting have been relatively well identified, with the support of lighting professionals. There are, however, still some controversies, such as those about its effects on security (Atkins et al., 1991; Mosser, 2007). On the other hand, the analysis of the values of the night resource remains a relatively unexplored field, with many questions about modes of valuation and its commensurability, measure, aggregation and generalisation. There remains an even more complex question about comparing the benefits form artificial light and the night sky. The question is framed in the context of ecosystem services assessment in that it involves many divergences among beneficiaries' and stakeholders' views and motivations for their participation in any protection initiatives (Burkhard et al., 2012; Chan et al., 2012).

Finally, we can say that light pollution raises questions about the value of services provided by the night and the location at which benefits are received. We want to demonstrate how this process is identifiable in the emergence of policy and planning initiatives that valorise night skies, especially for tourism activities. The night is no longer a passive frame to be removed or enhanced by lighting. It also provides a range of services and benefits appreciated by many people, particularly those linked to a "back-to-nature" dynamic, and offers new opportunities for rural areas.

Emerging and geographically localised services around the night: from a new consideration of night activities to new opportunities

Nighttime is a particular space-time, which contributes to building and shaping the relations of our societies to their environment. In Western civilisation, our populations have a complex relationship with darkness. Darkness can both reveal anxieties and fears and encourage imagination and creativity (in art, sciences, etc.). Light, the antithesis of the black and the night, traditionally performs symbolic work to convey the values of goodness, faith, truth and knowledge. Historically, the gradual rise of artificial light – especially of public lighting – is associated with the spread of progress, a part of the project of modernity. It thus follows the development of large networks (gas in the 19th century and electricity between the late 19th century and the 20th century). The emergence of urban lighting in the industrial era has also allowed the development of a wide range of activities at night, whether productive or leisure, and recreation activities. Daytime life is prolonged, albeit in a different setting (physically and symbolically), but with activities that are not inherently related to the night or based on its value. Instead, they often require an injection of artificial light. We can mention festive activities (bars, restaurants, concerts, etc.) or touristic enhancements of the city at night (festival of lights, sound and light, night markets). In these *overnight* activities, darkness is at best considered as a black screen, a frame for projecting light and living our sleepless nights. Nevertheless, the value of light festivals (e.g. La Fête des Lumières in Lyon, France) is related to the spectacular properties of artificial light, but it has a visceral need of a certain quality of shadows to be deployed with nuance. Beyond these events, some lighting professionals today call for less 'enlightening' projects, to avoid light competition and the excessive touristification of cities. We can see that now although nighttime is considered as extension time for traditional economic activities, a social demand exists – even in urban areas – for more subtle lighting and a larger role for shadow and darkness. This demand joins with concerns over light pollution discussed in the previous section.

At the same time, in several low-density residential areas, we can observe the development of activities *of the night*. Here, the quasi-absence of urban lighting constitutes a boon for activities built on the astronomical and biophysical definition of the night as a time of natural darkness and starry skies. Examples of such *of the night* activities are largely related to tourism: nights under the stars, bivouac (in mountains, deserts), astronomical events (star parties, such as Night of the Stars in France; see Figure 10.1), the development of astrotourism (Charlier

Figure 10.1 See the Milky Way in Yosemite National Park. Poster created by Tyler Nordgren as part of a North American National Park campaign

and Bourgeois, 2013) and ecotours or hikes to observe and listen to the night and its wildlife.

Figure 10.2 demonstrates our effort to describe the continuum between *overnight* and *of the night* activities. The diagram is built around major dichotomies that structure the different uses of the nocturnal time and resources. We consider these uses as resulting from the combination of different psychosocial factors, different values projected in the activities and different relations between people, night, darkness and artificial light. Thus, this conceptual and heuristic model, made from our observations and field experiences, allows us to highlight socio-economic determinants (for *overnight* activities) and environmental determinants (for activities *of the night*). Our work showed that *in the night* activities value mainly socialisation processes, the extension of daytime and economic activities lighting technical artefacts (e.g. buildings). The activities *of the night*, on the other hand, refer to the contact with "rediscovered nature" or with a kind of "lost identity" (Berque et al., 1994), through the darkness and the night sky. From this point of view, artificial light creates a de facto impairment that reduces access to "natural" objects.

Three examples can illustrate the *overnight–of the night* continuum. Light festivals are mainly social events, aiming to bring people together around an ephemeral change of the urban landscape. Night is important, but in its role as a frame to allow the enjoyment of different lighting practices, designed by inhabitants and lighting professionals. At the opposite side of the continuum, we find astrotourism. This activity requires as much natural darkness as possible to be able to enjoy stargazing and nightscapes. Hence, the most valued locations for astrotourism are generally remote areas, free from human interference (but also from natural restrictions, like clouds). Astrotourism can provide a double reconnection with nature and its values. First of all, it reconnects through the enjoyment of a natural object, the starry sky, and the experience of the relativity of human life it provides. Second, because it requires a preserved – or restored – nature, it pushes practitioners towards a new relation with nature and a stronger commitment to preserved space of naturalness (like other activities, climbing for example). Nevertheless, certain forms of astrotourism are at the same time activities that require costly high-tech artefacts. Between the two poles of the continuum, there is a wide range of activities, in which positions can vary, even for the same kind of activities. For example, hiking during the night could emphasise the sportive aspect, the night being a kind of original scenery for a commonly daytime activity. Or it could be oriented more towards connecting people with night wildlife, its "proper" landscape (including the starry sky), its wild noises and other "authentic" experiences of nature. In the first case, the value is not directly connecting with the night and the nature during nighttime, unlike the second case.

Recreational activities *of the night*, which have as their central object the observation or contemplation of what we collectively refer to as "nocturnal environmental resources" (nocturnal landscapes, starry sky, night sky phenomena but also sounds of nocturnal wildlife, among others; see RENOIR Research Group, 2014), strongly mobilise the idea (or myth) of a return to a "pristine nature".

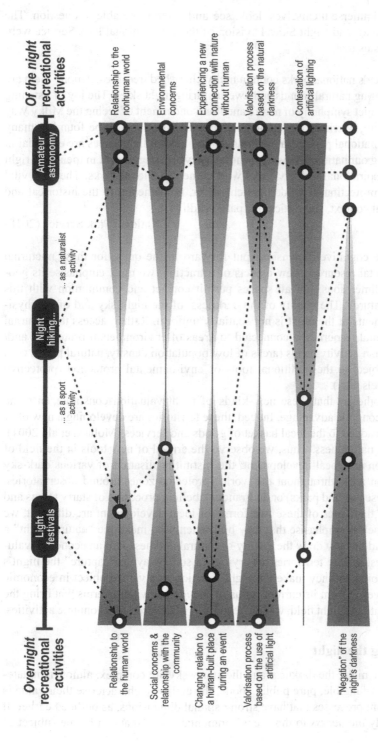

Figure 10.2 Conceptual diagram situating different activities in the continuum between *overnight* and *of the night* activities

Source: RENOIR Research Group (2014)

Spectators immerse themselves, look, see and listen to be able to question. The Natural Sounds and Night Skies Division of the U.S. National Park Service website informs us that

> America's national parks contain many cherished treasures. Among them are captivating natural sounds and awe-inspiring night skies. The joy of listening to the quiet symphony of nature and the wonderment of seeing the Milky Way stretching overhead are unique experiences that can still be found in many of our national parks. Natural sounds and natural lightscapes are essential in keeping our national treasures whole. They are magnificent in their own right and inspirational to the visitors who come to national parks. They are vital to the protection of wilderness character, fundamental to the historical and cultural context, and critical for park wildlife.
>
> National Park Service (2018)

Beyond the cognitive dimensions put forward in the quest for these nocturnal environmental resources, new objects of attraction, we must emphasise its geographical dimensions: not all spaces permit contact and communion with this "pristine nature". The quality of the darkness, of the night sky and of biophysical components of the night is not spatially uniform. Rather, access to nocturnal environmental resources is connected to areas of environmental protection and/or low human activity levels (areas of low population density, natural parks, etc.) and are subject to the traditional tools of environmental protection (protective zoning, labels, etc.).

We hypothesise that these new kinds of night valuation constitute, in some areas, an economic advantage. Indeed, these territories are developing a new offer that will be added to the local basket of goods and services (Mollard et al., 2001), focused on naturalness. Thus, we observe the arrival of new tools in the field of conservation and local development: the institutionalisation of various dark-sky preservation areas throughout the world (protected zones around observatories, dark-sky reserves and parks) or, in France, labelling processes of starry towns and villages. If the logics of these two forms of local development are different, we can nevertheless emphasise that they both attempt to make the "natural night" a resource and thereby draw the quality of natural darkness into an economic valuation process. These territories are trying, in some way, to "capture" the night's value. In doing so, they make the night function as a valued object in economic terms, a reversal from its current status (*overnight* activities), terms that bring the economy into the night field, where the night is a backdrop for economic activities.

Protecting the night

Because the night, the darkness and the starry sky are complex, almost nonmaterial and non-fungible, pure public goods, the ability to characterise their value is derived from processes that have strong spatial dimensions, as outlined earlier. It is especially the *access* to those environmental goods that can be the subject of a valuation process. Protected zones around observatories, dark-sky reserves and parks or other labelling processes promote privileged access to the night and different environmental goods and services associated to it.

Certification initiatives for dark-sky protected areas (DSPA) have grown worldwide. Since the early 2000s, several national organisations have worked to create their dark-sky preserves programmes: the Royal Astronomical Society of Canada, the International Dark-Sky Association (IDA, USA) and, in 2007, the StarLight Initiative was founded, in partnership with UNESCO and the International Astronomical Union as: "an international action in defence of the values associated with the night sky and the general right to observe the stars . . . open to the participation of all scientific, cultural, environmental, and citizens' organizations" (Starlight, 2019).

These initiatives stress an increasing concern about the problems of light pollution caused by socioeconomic activities, the functioning of infrastructures and urban sprawl. The International Dark-Sky Reserve (IDSR), certified by the IDA, is the most demanding label, because it goes beyond the protection of the existing capacity of nightlife. It involves a goal of improving the quality of the night sky and therefore requires technical adjustments and monitoring policies for the quality of the night sky and nighttime wildlife. Currently located in North America (the majority of DSPAs are located in the US or Canada), they are reaching Europe, where numerous movements for creating DSPAs have arisen (Beskydy, the first dark park designated in 2013 at the border of Slovakia and Czechia; Northumberland in the UK; Pic du Midi de Bigorre in the French Pyrenees, for example). Beyond the problems of light pollution and its regulation, these initiatives stress other topics regarding geographical dynamics. Their arrival in more-inhabited areas, like Western European countries, point out new ways of considering the (nocturnal) environment.

Indeed, we move from a model focusing on local and specific protection (mainly for astronomical observatories) to a more global and generic protection (Depraz, 2008), which signifies a more comprehensive consideration of the night, one that is based on all services associated with it (culture, environment, health, etc.). This shift towards a more general concern about environmental protection is also observed in the governance and strategies of night and starry sky protection associations (Challéat and Lapostolle, 2014; Meier, 2015). A key consequence is that night protection is now part of environmental policy at different scales but also that the night is newly considered as an environmental asset, one that could increase the naturalness value offered by *territoires*[2] and be integrated into the local bundle of environmental goods and services.

We propose that the night has become a local resource and aim to stress the different ways people and *territoires* use this resource – in particular in DSPAs – and to understand the specificity of these resources. The night sky benefits a set of economic activities, and in this way, it generates added value. From the point of view of local touristic offers, it generates new niche activities (like stargazing and astrotourism), and it strengthens and revalues existing activities (e.g. nature-based tourism) by integrating a nightscape dimension. It is through these activities that the night sky resource is activated, converted into monetary value and appropriated by local actors. We can observe this trend through communication strategies developed by tourist offices based on the night sky and the emergence of new activities (tours to observe night sky, new kinds of accommodation, night hiking to observe and listen to the night life, etc.). For example, in the Mont Mégantic International Dark-Sky Reserve, new hotel business advertised Starry Sky EXP Chalets, described as

> Original architecture, comfort and intimate contact with nature combine to ensure an unforgettable experience. Imagine a well-glazed home, letting nature permeate the interior and giving you the impression of always being a little outside. . . . With this type of accommodation, the park offers an immersion from the Earth to the stars, with proximity to the forest, to a watercourse and to the starry sky of the first International Dark-Sky Reserve.[3]
>
> (authors' translation)

Beyond these direct benefits to tourism, a preserved night sky can consolidate a distinctive place-identity, can enhance community well-being and may contribute to a region's cultural heritage by working to develop a sense of belonging. As the Estrie Observatory for Community Development's website enthuses,

> If you have chosen to live in Chartierville, Hampden, La Patrie, or Lingwick Scotstown, you can be proud to be at the heart of Canada's first globally recognised and inhabited dark-sky reserve. Located at the foot of the majestic Mont Mégantic and the prestigious astronomical observatory of the same name. The nature and purity of heaven as a living environment. Not bad, right?[4]
>
> (authors' translation)

This shows that non-monetary values may play a significant role. Among these, heritage value plays a key role in the appropriation process, or at least in the increase of "concernment" (Brunet, 2008) that could lead to commitment.[5] A recent episode in the history of the Mont Mégantic Observatory (MMO), located in the heart of the Mont Mégantic National Park and the IDSR of the same name, perfectly illustrates these processes.

On the morning of 11 February 2015, the MMO, managed by the University of Montreal, announced its closure on 1 April 2015 for budgetary reasons following the non-renewal of a federal grant. This announcement was quickly relayed via social networks and raised unexpected large-scale reactions: many people mobilised for maintaining the observatory activities. The next day, the federal government reversed its decision: Ottawa announced funding for two years to keep the observatory activity. In an interview in June 2015 with two astrophysicists of the MMO, the important role of the Mont Mégantic IDSR in this "rescue" appears. Indeed, this IDSR is strongly associated in the collective imagination with the presence of the observatory in its centre. Without this amalgam between IDSR and MMO, the observatory's image would not be that of a public good with a heritage value (Box 1). It is because of this value, revealed through the emergence of an IDSR, that it was possible for the Canadian federal government to reconsider its willingness to pay. At the same time, the presence of the IDSR induces the integration into the local identity of new considerations about night skies and the different related objects and activities. Another example is the development of education practices around the night sky (citizen science, the renaming of a school, etc.). Finally, the event around MMO illustrates a process of value activation and of the appropriation of nocturnal environment resources.

Box 1   Extract of an interview with two astrophysicists of the Mont Megantic Observatory (2015, translated from French)

- OMM 1: We tried to have other funding for the operation of the observatory but we never did, so we decided to close down on the 1st of April, and this is when the federal government decided to fund for two years what was missing on one-time grant. There are federal elections in October this year. . . .
- OMM 2: There are elections, and I have to admit it is also the media pressure.
- OMM 1: Yes, it was very, very strong! I deal with social networks, and I simply couldn't keep up at all, I had my Twitter account scrolling all the time!
- OMM 2: It had been known around the world! Half an hour after the announcement, I received a call from someone in Dubai who wanted to know what he could do to help us!
- OMM 1: A call for donations had been done, crowdfunding, quickly, and immediately we calmed saying, "wait, wait, let us pull ourselves together!"
- OMM 2: And this is a very fine example of this kind of match between the observatory and recreational and tourism activities. We have nothing to do with it, but it's because there is an IDSR that people know the MMO is there and that there is no discussion of closing it! For us, it has been beneficial in the end! We spoke earlier of cohabitation between tourism and astronomical activities. Does it go well? . . . But there is no choice! Our survival depends on this now!
- OMM 1: I must say that in Quebec people are very proud and quickly appropriated things. And with the observatory, and thanks to the work of Hubert Reeves in all media, people have appropriated sentimentally the observatory, it is something important for them, really.
- OMM 2: It has become a national treasure, and you can't affect a national treasure! You do not affect the observatory!

Box 1 continued

- OMM 1: And when there was the announcement, people who had absolutely nothing to do with this field then called me and asked me what to do, how they could intervene in order to save the observatory.
- OMM 2: The reaction was extremely strong in the region because as liked to say the former prefect of a municipality, "the observatory is the flagship". The rest is dependent on the observatory, so if you close it. . . . They feel that the magnet passes through the observatory, even if everything – the International Dark-Sky Reserve, the Astrolab – will continue to exist if the observatory disappears. But it is the symbol! So there has been a very strong reaction in the region, but it's beyond the scope of the region; it was all over the province and even in the rest of Canada.
- Question: The Observatory and the sky above are becoming a heritage?
- OMM 2: Yes, absolutely! Absolutely! Exactly! And I would say it's the way the product is sold, isn't it? Without being pejorative, that's how it is marketed! "Come and see the stars, come and see the observatory!"
- OMM 1: There is this belonging; people like that and it persists. Well, the costs to maintain it are astronomical anyway, but . . .

Source: Interview on 8 June 2015, in Montreal, Quebec

It appears that the (re)valuation process concerns all ecosystem services and benefits connected with a preserved dark night sky. Professionals and politicians are appropriating this process to activate new resources for their *territoire*, through different innovative devices and with several objectives. Indeed, the idea of the night sky as resource refers to technical, organisational and institutional innovations, designed to advocate for its preservation and valorisation. A DSPA enables new forms of public action, which re-mobilise existing networks of actors and cultivate new ones. It also implies new technical practices and new ways to manage them. This is the case, for example, with new lightning technologies (type of light sources, direction, smart lighting) and new governance for public lighting systems. New laws and norms are elaborated, but we also observe the inclusion of new stakeholders in the decision process – for example, regarding the possibility of switching off the light in certain places and/or at a certain time during the night. From this point of view, DSPA actors experiment with new technical, organisational and institutional devices that could be a reference for other rural or urban areas. Indeed, through their action to valorise and protect the NS, as a territorial resource, they reveal and activate monetary and non-monetary values. At the same time, they propose new ways to balance night skies and artificial lighting benefits and drawbacks.

Conclusion

Our consideration of the value of the night here applies mainly to its starry sky, because of its function as a resource for emerging niche tourism activities. The night sky is emblematic of a set of services, benefits and values associated with nighttime and natural darkness; indeed, it serves as a reference to measure night quality (e.g. through the visibility of the Milky Way). Revealed by the excess of light and in a context of an increasing preoccupation for the environment, night sky values are now widely recognised. They are implicitly or explicitly integrated into a set of activities that emphasise wild night. A preserved night sky is considered a local resource for *territoires*, and new actions and devices have been emerging to reveal and activate this resource. This is the case of Dark-Sky Protection Area certification. This leads to new trades-off in favour of night skies versus artificial lighting, considered a source of pollution. Considering the night sky as a resource incorporates it into capitalism via a form of hypertourism, the touristification of an otherwise "pure" space-time. Nevertheless, one can also consider protection of this resource as a shift towards an ecologisation of local development strategies. (Challéat et al., 2018).

This logic seems to diffuse into other spaces, beyond the heart of DSPAs. Around DSPA initiatives, relations between urban and rural are no longer asymmetric. Nocturnal environment preservation is also a territorial resource for urban areas. In this way, DSPAs highlight complex relations between landlocked areas and urban cities. They play a retroactive role in land and urban planning. In the case of the DSPA of Pic du Midi de Bigorre (France), agreements have been concluded between the reserve and surrounding towns (Lourdes, Pau, Tarbes), to reduce light pollution (e.g. the installation of less-polluting and more-energy-efficient lighting equipment during renovation or lamp replacements). In the background of these agreements is a concern to reduce the financial burden of lighting and its energy use, which strengthens the argument related to the services provided by the night sky (whose own value might be less likely to be revealed, appropriated and mobilised by public authorities, without the clincher of cost savings). Economic aspects appear to have a strong influence in the trade-off between the benefits of artificial lighting and a preserved night sky.

Nevertheless, the question remains whether this logic can be applied to the environmental impacts of artificial lighting and even more to health impacts. Indeed, the trade-off between artificial light and the dark night has different characteristics and results depending on the location (urban vs rural). The contradiction between two public goods, natural night and artificial lighting, requires more-detailed and interdisciplinary investigation, comparing different type of territories in different national contexts. Even if the commensurability problem of valuing and protecting the night has no clear technical or economical solution, political and social concerns illustrate new relations between society and nature that are potentially relevant to the wider ecosystem services debate.

Finally, the increase in initiatives to protect the dark night and its starry sky should not overshadow the plurality of the night. The various "physical" definitions of this period of darkness show themselves: before the dark night, several twilights follow each other. Depending on the angle of the sun below the horizon, the twilight is civil, nautical or astronomical. At each of these stages, it deploys a particular brightness in the sky: this is not the day, but it is not yet dark night. These moments of transitions (sunrise and sunset, moonrise and moonset, the appearance in the sky of the first planets and bright stars, etc.) are particularly sought after by astrotourists and contemplatives: they reveal unique landscapes and have a great emotional charge. The perception of the night is itself complex because it is highly individualised and dependent on what one comes to seek in this space-time. Deep-sky astronomy (observing star clusters, nebulae, galaxies) requires the night sky to be as dark as possible; here the light reflected from the Moon becomes a source of annoyance. At the same time, moonlight illuminates the scenery and is one of the interests of the night for contemplatives or landscape photographers, for example.

The difficulties in the economic assessment of *the night* are due to the inherent complexity of this space-time and its perception, as much as the range of configurations encountered in tools and processes that intend to manage or to protect and preserve.

Notes

1 See in this regard the debates on international treaties governing property rights in outer space: Husby (1994), Baca (1992) and Gorove (1974). For more recent examples, see Sattler (2005) and Gaspari and Oliva (2019).
2 We agree with Barreteau and colleagues' (2016) definition of *territoire*. Within francophone social sciences, this does not translate to the term "territory", which is associated with administrative boundaries. Rather, *territoire* is defined as "a spatial mediator of all social life" (Di Méo, 1999). *Territoire* is both a social and a lived space, including political and ideological dimensions of space. *Territoire* "emerges through the interactions that individuals and groups have with their environment within a specific geographical area" (Debarbieux, 2007).
3 www.tourism-megantic.com. 2018. Parc national du Mont-Mégantic | Tourisme Mégantic [online]. Available at www.tourisme-megantic.com/fr/dormir/parc-national-du-mont-megantic. [Accessed 20 November 2018].
4 www.oedc.qc.ca. 2018. Fiche synthèse de la communauté de la Zone Est | oedc.qc.ca. [online]. Available at www.oedc.qc.ca/tableau-de-bord/06/haut-saint-francois/zone-est. [Accessed 20 November 2018].
5 The concept of concernment describes a posture that "is never fully resolved between passivity and activity, it remains at least potentially active. It lies in an in-between that which can, in the presence of any meaningful event, turn into mobilization" (Brunet, 2008; authors' translation).

References

Atkins, S., Husain, S. and Storey, A. (1991). *The Influence of Street Lighting on Crime and Fear of Crime*, Crime Prevention Unit Paper, no 28. London: Home Office.
Baca, K.A. (1992). 'Property Rights in Outer Space', *Journal of Air Law and Commerce*, 58(4), pp. 1041–1085.
Barreteau, O., Giband, D., Schoon, M., Cerceau, J., DeClerck, F., Ghiotti, S., James, T., Masterson, V.A., Mathevet, R., Rode, S., Ricci, F. and Therville, C. (2016) 'Bringing Together Social-ecological System and Territoire Concepts to Explore Nature-Society Dynamics', *Ecology and Society*, 21(4).
Berque, A., Conan, M., Donadieu, P. and Lassus Roger, A. (ed.) (1994). *Cinq propositions pour une théorie du paysage*. Paris: Champ Vallon.

Besecke, A. and Hänsch, R. (2015) 'Residents' Perception of Light and Darkness', in Meier, J. et al. (ed) *Urban Lighting, Light Pollution and Society*, pp. 224–248.

Brown, T.C. (1984) 'The concept of value in resource allocation', *Land economics* 60(3), pp. 231–246.

Brunet, P. (2008) 'De l'usage Raisonné de la Notion de "Concernement": Mobilisations Locales à Propos de l'industrie Nucléaire', *Natures Sciences Sociétés*, 16, pp. 317–325.

Burkhard, B., Kroll, F., Nedkov, S. and Müller, F. (2012) 'Mapping Ecosystem Service Supply, Demand and Budgets', *Ecological Indicators*, 21, pp. 17–29.

Challéat, S. (2010) *"Sauver la Nuit" – Empreinte lumineuse, urbanisme et gouvernance des territoires*. Unpublished doctoral thesis, Université de Bourgogne, Dijon.

Challéat, S. and Lapostolle, D. (2014) '(Ré)concilier éclairage urbain et environnement nocturne: les enjeux d'une controverse sociotechnique', *Natures Sciences Sociétés*, 22(4), pp. 317–328.

Challéat, S. and Lapostolle, D. (2018) translated by Waine O. *Getting Night Lighting Right. Taking Account of Nocturnal Urban Uses for Better-Lit Cities*. Metropolitics, 2 November 2018. Available at: www.metropolitiques.eu/Getting-Night-Lighting-Right.html.

Challéat, S., Lapostolle D. and Milian, J. (2018) 'The Night-Time Environment in French Mountain Areas. A Resource and a Transition Operator Towards Sustainability', *Journal of Alpine Research | Revue de géographie alpine* [Online], 106(1): http://journals.openedition.org/rga/3947

Chan, K.M.A., Satterfield, T. and Goldstein, J. (2012) 'Rethinking Ecosystem Services to Better Address and Navigate Cultural Values', *Ecological Economics*, 74, pp. 8–18.

Charlier, B. and Bourgeois, N. (2013) ' "Half the park is after dark" – Les parcs et réserves de ciel étoilé: nouveaux concepts et outils de patrimonialisation de la nature', *L'Espace Géographique*, 42(3), pp. 200–212.

Debarbieux, B. (2007) Territoire-territorialité-territorialisation: aujourd'hui encore, et bien moins que demain, in Vanier M. (ed) *Territoires, territorialité, territorialisation: controverses et perspectives*, Rennes, France: Presses Universitaires de Rennes.

Depraz, S. (2008) *Géographie des espaces naturels protégés*. Paris: Armand Colin.

De St-Exupéry, A. (1943) *The Little Prince*. New York: Reynal & Hitchcock.

Di Méo, G. (1999) 'Géographies tranquilles du quotidien. Une analyse de la contribution des sciences sociales et de la géographie à l'étude des pratiques spatiales', *Cahiers de Géographie du Québec*, 43(118), pp. 75–93.

Dupuy, P.O., Girard, F., Challéat, S., Bénos, R., Lapostolle, D., Poméon, T. and Milian, J. (2015) 'The Role of the Northern Lights in the Production of a New Touristic Imaginary of the Nordic', *The Dynamics of Darkness in the North*, February 26–28, 2015, Reykjavík, Iceland.

Edensor, T. (2015) 'The Gloomy City: Rethinking the Relationship Between Light and Dark', *Urban Studies*, 52(3), pp. 422–438.

Ernstson, H. and Sörlin, S. (2013) 'Ecosystem Services as Technology of Globalization: On Articulating Values in Urban Nature', *Ecological Economics*, 86, pp. 274–284.

Gallaway, T. (2010) 'On Light Pollution, Passive Pleasures, and the Instrumental Value of Beauty', *Journal of Economic Issues*, 44, pp. 71–88.

Gallaway, T. (2015) 'The Value of Night Sky', in Meier, J. et al. (ed) *Urban Lighting, Light Pollution and Society*, London: Routledge, pp. 267–283.

Gallaway, T., Olsen, R.N. and Mitchell, D.M. (2010) 'The Economics of Global Light Pollution', *Ecological Economics*, 69, pp. 658–665.

Gaspari, F. and Oliva, A. (2019) 'The Consolidation of the Five UN Space Treaties Into One Comprehensive and Modernized Law of Outer Space Convention: Toward a Global Space Organization', in *The Space Treaties at Crossroads*, Cham: Springer, pp. 183–197.

Gorove, S. (1974) 'Property Rights in Outer Space: Focus on the Proposed Moon Treaty', *Journal of Space Law*, 2(1), pp. 27–80.

Husby, E. (1994) 'Sovereignty and Property Rights in Outer Space', *Journal of International Law and Practice*, 3, p. 359.

Kull, C., Arnauld de Sartre, X., Castro-Larranaga, M. (2015) 'The Political Ecology of Ecosystem Services', *Geoforum*, 61, pp. 122–134.

Marín, C. and Jafari, J. (eds) (2007) *Starlight: A Common Heritage. International Initiative in Defence of the Quality of the Night Sky and the Right to Observe the Stars*, Canary Islands, Spain: Starlight Initiative and Instituto de Astrofisica de Canarias.

Meier, J. (2015) 'Designating Dark Sky Areas', in Meier, J. et al. (ed) *Urban Lighting, Light Pollution and Society*, London: Routledge, pp. 177–196

Millennium Ecosystem Assessment. (2005) *Ecosystems and Human Well-being: Synthesis*. Washington, DC Island Press.

Mollard, A., Pecqueur, B. and Lacroix, A. (2001) 'A Meeting Between Quality and Territorialism: The Rent Theory Reviewed in the Context of Territorial Development, With Reference to French Examples', *International Journal of Sustainable Development*, 4(4), pp. 368–391.

Mosser, S. (2007) 'Éclairage et sécurité en ville: l'état des savoirs', *Déviance et Société*, 31, pp. 77–100.

Narisada, K. and Schreuder, D. (2004) *Light Pollution Handbook*. Astrophysics and Space Science Library, Berlin/New-York/London, Springer.

National Park Service (2018) Natural Sounds and Night Sky Division. Available at: https://www.nps.gov/orgs/1050/index.htm.

Ostrom, V. and Ostrom, E. (1977) 'Public Goods and Public Choices', in Emanuel, S S. (ed) *Alternatives for Delivering Public Services: Toward Improved Performance*. Boulder, CO: Westview Press, pp. 7–49.

RENOIR Research Group. (2014). See RENOIR website. Available at: http://renoir.hypotheses.org.

Riegel, K.W. (1973) 'Light Pollution. Outdoor Lighting Is a Growing Threat to Astronomy', *Science*, 30, (179), 4080, pp. 1285–1291.

Sattler, R. (2005) 'Transporting a Legal System for Property Rights: From the Earth to the Stars', *Chinese Journal of International Law*, 6, 23.

Simpson, S.N. and Hanna, B.G. (2010) 'Willingness to Pay for a Clear Night Sky: Use of the Contingent Valuation Method', *Applied Economics Letters*, 17, pp. 1095–1103.

Starlight (2019) Objectives of the Starlight Initiative, Starlight. Available at: https://www.starlight2007.net/index_option_com_content_view_article_id_199_objectives_catid_62_the-initiative_itemid_81_lang_en.html.

Willis, K. (2015) 'Improved Visibility of the Night Sky', in Meier, J. et al. (ed) *Urban Lighting, Light Pollution and Society*, London: Routledge, pp. 249–266.

11 Value and diminishment

Listing state park closures, the 2011 attempt to meet general fund reductions in California

Gareth Hoskins

[T]his governor has dragged us into the state budget fight that the nurses, the teachers, the doctors and everybody else had been fighting for decades, so welcome to the fight, the state budget fight and I think that this governor has laid down a gauntlet and said, "okay, figure out a solution" and Nevada County is at the front edge of figuring out that solution.

(Chief Executive Office Sierra Fund,
12 December 2011, town hall meeting)

In March 2011 the California State Legislature adopted Governor Gerry Brown's proposed budget that reduced contributions to the state park system from the General Fund by US$11 million in that fiscal year (2011–2012) and another US$11 million the next. The Parks Department, having faced years of cuts, escalating expenses and the recent public rejection of a financial rescue package, responded by saying it would permanently close 70 parks, a quarter of the department's holdings. They had three weeks to draw up a list of closures and published the list in a press release, along with a series of "factors considered" in devising the list on 13 May that year.

My focus in this chapter is on the closure list, the tensions surrounding its compilation and what these might tell us about the operation and deployment of value in the public sphere. The chapter does not advocate for the value of parks per se but rather examines the various mechanisms by which value is located because, I argue, it is this very endeavour to locate value that reveals something about the nature and operation of value more generally.

The functional legitimacy of the preservation sector is supported by two value-related orthodoxies: an ontological orthodoxy (i.e. a belief in the existence of value itself as an intrinsic, essential property) and an epistemological orthodoxy (i.e. a belief in the possibility of value's precise and accurate identification). In more applied terms, the practice of listing and delisting areas of land as parks or natural monuments relies on an acceptance that some places are more worthy of esteem than others and that superiority is detectable and, indeed, measurable in gradations.

This chapter aims to undermine our confidence in these kinds of geographical valuations and the designations they proliferate because they work to roll out a version of the world that supports the aesthetic preferences of the liberal elite while claiming to work for a higher, less-self-serving purpose. I use Malakoff Diggins State Historic Park in Nevada County California, one of the parks on the 2011 closure list to explain how value is relational and becomes subject to efforts of fixing. The chapter works through what I have previously called a geographical axiology (Hoskins, 2016) first to ask that we more fully acknowledge the radical contingency of value in formal decisions about what parts of the earth are praised and protected and second to illustrate how the practice of valuation functions to redefine "public" as a category in a context of fiscal crisis. What kinds of public emerge when fine-grained negotiations about revenue streams and budgetary concerns come up against broader social ideals and long-held principals about the value of public parks and the treasures they contain? As a normative judgement upheld by a range of institutional logics and technologies, geographical value, in the same way as any other kind of value, is produced through the act of its measurement. If we accept that intrinsic value is an illusion – that it is made rather than found – then we must think more carefully about how its identification works to serve the interests of those who support and sponsor it.

The logics of devaluation

Although couched in reassuring language of regret and encouraging statistics about revenue and visitation, when published at the start of the 2011 summer, the California State Parks closure list provoked an outcry. Letters, protests and community crisis meetings took place; legislators were picketed at the state capitol building in Sacramento by school children bused in to hand over petitions asking why their backyard, their playground, their classroom was being shuttered.

In a public meeting held to raise funds and organise resistance, the executive director of the South Yuba River Citizens League, the main advocacy group leading the campaign, mobilised a certain set of values around nature and the public's right to gain access:

> We're not going to stand by now and allow these treasured parks with their wonderful swimming holes, their beautiful trails, to be abandoned by the state. We're here tonight, all of us in this room, with one shared purpose, to prevent our state parks from closure. I say to Governor Brown, Resource Secretary Laird and Parks Director Coleman, remove these state parks from the closure list!
>
> (Dartik, 2011)

State Republicans joined the rancour, saying the parks department were putting their dysfunctional bureaucracy ahead of the recreational needs of hard-working families, and the State Assembly's Accountability and Administrative Review Committee held a special hearing to force the parks department to explain the list. Its appointed chairman said "We're the 'show your math' committee" and told the press afterwards that the process was "beginning to seem downright arbitrary and capricious in the absence of any data. . . . I was not satisfied to learn that the park closures process consisted of 12 unnamed people in a room throwing darts at a wall" (Weiser, 2011).

The parks department responded by saying,

> there was a great deal of argument and arm wrestling, [but] there wasn't really a rubric. . . . we had to rely on the experience of our professional staff . . . We had 12 long-employed highly trained park superintendents who knew the parks personally and individually. Do we like closing Parks? No! But the choice was between that and reducing services everywhere to an unacceptable level. We thought about developing a scorecard, but there are so many complexities that we gravitated toward experience, judgement and narrative analysis. This is not a golf game.
>
> (Renda, 2011)

It's not the first time sporting metaphors appeared in the arena of public policy, but their use here hints at the rarely acknowledged zero-sum nature of valuation as an administrative technology. Just as in golf, darts and arm wrestling, where one succeeds at the expense of another, so it is in the distribution of funding.

Legislators were trying to steer the debate away from the value of parks per se to a debate concerning procedural misdemeanour in the ranking of one park's value against another. Here we see the denial of equivalence and the performance of valuation as a necessary action that the state can ensure is carried out with probity.

When we value something, we are never simply saying that one thing is good; we are unavoidably saying that one thing is better than another, one place should be ranked higher than another, one application for funding is more worthy than another and so on. In his analysis of the value concept as "a conceptual virus spread by modernity", Clarke points to Baudrillard's (1998) essay *The Destiny of Value* that declares: all that lives by value will perish by equivalence "because we no longer know what is true or false, what is good or evil, what has value or does not, we are forced to store everything, record everything, conserve everything, and from this an irrevocable devaluation ensues" (Clarke, 2009, p. 4).

Analytical focus on the performance of value – that is, valuation as a practice rather than as an object or foundational essence – helps to move the conversation beyond what gets counted as valuable and what doesn't towards considering valuation itself as a means of getting things done. When public institutions are forced to champion their impacts on society and celebrate their success in order to acquire funding, the zero-sum nature of the game is never acknowledged. The fact that any declaration of value involves the removal of value from something else is denied. "Value", like a host of terms central to the functioning of modernity, such as "the rational" or "the West", is defined by what it excludes. It has, in Derridean terms, a constituent outside. Value's Other is diminishment.

Axiological antagonism

The philosophical analysis of value has occupied an esteemed array of thinkers from philosophers Plato and Kant to Adam Smith and David Ricardo in classical economics; 19th-century thinkers, from Nietzsche (1972), Dewey (1922, 1939), Marx and Polyani (2001); and more recent scholars working in social anthropology, human geography, critical literary, aesthetics and poststructuralist theory. In this chapter, I draw on these theories to develop a relational axiology that asserts works to expose the power relations and methodological gymnastics required to maintain the appearance of value's legitimacy in an age where valuation (through audits, monitoring, performance management, targets and action plans) has become ever-more prevalent.

Human geographers have been thoroughly involved in the debates around value and its operation as a relational, power-laden concept. Jonathan Smith's (2002) essay on value and place in the United States goes as far as to assert that the entire discipline is largely about values. He adopts a relational position arguing that the value of a particular place is never absolute:

> No place is naturally valuable because its land is productive, its harbour deep, or its hills fled with gold; it becomes valuable only once it falls to the hands of a people who know how to make use of such things as fertile soil, a generous anchorage, or a precious metal. The value of any particular place depends on the culture of those who control it. The value of a valued place is always, in short, relational.
>
> (Smith, 2002, p. 53)

Other place-related value research in the discipline that recognises the relational conditions of value includes Gold and Burgess's (1982) *Valued Environments*, which maps the phenomenological and affective bonds between places, individuals and groups in the UK; Cresswell's (2012) account of archival practices that assign objects contrasting and competing value at Maxwell Street Market in Chicago, Illinois; and Miller's (2008) critique of bottom line thinking in the UK government audit culture of *best value*.

For some political economists, political ecologists, urban theorists and economic geographers, value has become a topic of study itself, which generates research questions as to how it is assigned, transferred and controlled and what value actually does in societies that hold to a dominantly capitalist mode of production (Soja, 1980; Harvey, 1982; Smith, 1984). Value is also an important means of theoretical positioning often used to crystallise disagreement about the proper status afforded to the economic as a totalising analytic. Roger Lee, for example, rejects the society/economy dualism that treats value as if autonomous and distinct from social relations. His account of an "ordinary economy" is one where value is historically and geographically variable and "emerges from the practice and performance of economic geographies" (Lee, 2006, p. 414).

Most relevant to the relational axiology I develop here, however, is research on the spatial politics of calculation and especially that which examines value's role in making nature more legible to the logics of private capital (Castree, 2008; Robertson and Wainwright, 2013; Sullivan, 2013; Yusoff, 2011). It is in this vein that I pursue my examination of the parks closures, but I also take inspiration from the research trajectories of those examining value in the heritage and preservation sectors. For example, National Parks historian John Sprinkle's (2014) account of how the criteria for assessing nominations to the U.S. National Register of Historic Places and National Historic Landmarks refers to awkward compromises, contradictions, ad hoc elisions, cover ups and instances of caving in to pressure that became necessary to ensure the criteria maintained a "veneer of academic objectivity".

Our current faith in value as an administratively useful concept has its roots in Enlightenment principles of perfectibility, where, more often than not, value is conflated with purity. "Purification" is the term Latour (1993) employs to describe the "modern" project of categorically separating humans from nature, and purity is implicit in Aristotle's theory of natural law, which holds that something is right if it fulfils its true purpose. Echoing this maxim, Robert Hartman's book *Structure of Value* posits his base principal of value with this phrase: a thing is good insofar as it exemplifies its concept. In other words, a thing is good if it possesses all the properties its name suggests that it should, if it fulfils its own definition, if it is pure. The contemporary application of this logic is perhaps most apparent in our treatment of mineral deposits that are graded according to the richness of the ore. We are familiar with its use in expressing the quality of racehorses (thoroughbred) or of Scotch whisky (single malt). The logic of purity works also in a preservation sector often driven by the phantom of authenticity. Historic sites must meet requirements of "integrity" if they are to communicate their story without too much distraction from the contemporary world, and "natural" sites are much more likely to be protected if they display elements that are characteristically pristine, unsullied and seemingly "unproduced".

The historian Richard White (2000) discusses the problem of our fixation with purity in relation to the environment:

> What many of us have done, myself included, is to find the root of many of the horrors of our time in a categorical mistake, confusing one thing with another, and in the transgressing of forbidden boundaries. We have logically deduced that the solution is purity: keeping categories separate, the boundaries intact.
>
> (p. 213)

Law, in his book *Organising Modernity*, highlights the danger of this:

> It seems to me that we have spawned a monster: the hope or the expectation that everything might be pure; the expectation that if everything were pure

then it would be better than it actually is; and we have concealed the reality that what is better for some is almost certainly worse for others; that what is better, simpler, purer, for a few, rests precariously and uncertainly upon the work and, very often, the pain and misery of others.

(Law 1993, pp. 6–7)

We see an expression of this in Hetherington's discussion of disposal at the museum in *Capitalism's Eye* (2008). It is also illustrated in relation to the heritage sector in what Harrison (2013) diagnoses as "the crisis of accumulation" – his term for the functional paralysis that comes with incremental expansion and associated mission-creep of listing and landmarking agencies that, for reasons of legitimacy, seek to draw in previously underserved communities with a wider appreciation of significance.

This takes us back to the list of state park closures that, for reasons only broadly sketched out, were deemed not to make the grade.

Value's inauguration

The three thousand-acre Malakoff Diggins State Historic Park, 26 miles from Nevada City California, was acquired by the park service in 1964 because of its association with the infamous practice of hydraulic gold mining in operation at a moment of rapid industrialisation in the post–gold rush era of the western United States.

The official opening ceremony took place on Saturday 13 August 1966, conducted by the acting chief of the division of parks and beaches Earl Hanson, who arrived in a 1924 Dodge owned by Charlie Gauss, the last surviving hydraulic miner to work on what was once the largest hydraulic gold mining operation in the United States. Hanson took the platform with state senators, city councillors and district assembly members who performed a rendition of Woody Guthrie's "This Land Is Your Land" sung with around a hundred of the town's past residents.

Back in 1967, there were grand ambitions to open a period restaurant and hotel with horse drawn transportation throughout and provisions for a thousand campers. The interpretive Services Division had annual projections of 250,000 visitors and 100,000 dollars in revenue, which was thought to be a conservative estimate (CDPR, 1967 Master Plan Narrative, p. 2).

By then, the high gravel cliffs of the Malakoff pit exposed by hydraulic hoses carving through the mountain in search for gold had gained notoriety as for its visual appeal and geological curio. The Mineral Information Service (a California monthly news release published by the Department of Natural Resources) described its visual value in a glossy promotional publication accompanied by large format photographs:

The enormous pit has, with the passing of the years become one of the state's precious scenic resources – a miniature Bryce Canyon set in a dusky evergreen forest, with the little old town of North Bloomfield drowsing quietly in filtered sunlight near its eastern edge. It may be difficult to grasp, in the quiet of today, the tremendous fury of activity with which men inadvertently created this lovely spot. The miners are gone, and the giants no longer inhabit this part of the earth: but they have left their mark, carving in a few short years beauty that nature measures in the long eons of geologic time.

(Egenhoff, 1965, p. 47)

The visual value is reaffirmed by the site's adoption into the National Register of Historic Places in 1972 with the following justification:

The mining area is primarily an environmental resource, with a few trails lead-ing to overviews. Spectacular color lends picturesqueness to the crenellated spires which fringe the canyon, creating what amounts to a miniature Grand Canyon or Bryce Canyon. Here the California State Park System has set aside an area testifying to the irresistible force of water when played against banks of earth as much as 600 ft. in height. What has resulted is a monument that will be a unique attraction and abiding interest for years to come.

(National Park Service, National Register Inventory Nomination Form, 30 November 1972)

In the end, visitation was below initial expectations (it is now around 27,000 annu-ally). The park was too remote and inaccessible and there was too much gold rush heritage competition near population centres lower down the mountains. Revenue was below that anticipated and the operating costs were high. All-day electricity from a diesel generator was needed to maintain the temperature and humidity lev-els of historic buildings and artefacts and round-the-clock staffing to secure it all. For these reasons, the park had been threatened with closure four times already, and its presence on the 2011 list was of no particular surprise to the rangers.

But Nevada County residents complained bitterly when they saw Malakoff Diggins appear on the list. They tried to modify the state park's valuation, but the figures were difficult to access because operating budgets had evolved in ways that made things convoluted and difficult to untangle. A member of Malakoff Dig-gins Park Association described the situation in an email to me:

No one seems to know anything for sure about the status of Malakoff. Sev-eral people have tried to get some numbers from State Dept. of Parks & Rec. Either they cannot or will not tell us how much it costs to run the Park. (It seems to me that if they cannot produce those numbers with a press of one computer key, then they cannot say that closing the park will save any money – but I get caught up in conspiracy theories. . . . They are trying to hide the numbers from the public, or they are totally incompetent at book-keeping). Either explanation says that the decision of which parks to close, or to close any parks, is smoke and mirrors. Someone has a political solution for an economic problem.

(Park Advocate, 13 December 2011)

When figures did emerge, the situation looked even more suspicious:

> I started to dig down and find out that the state – and I'm not trying to be mean – the state is completely fraudulent with cost savings they originally posted for these parks. The parks, these numbers, which are public information, were 30–60% inflated. We're not going to save the money. Not only are we going to lose money to local economies and hundreds and millions of dollars, but the initial savings are not what they seem and that's just unfortunate.
>
> (Public land activist in radio broadcast, 13 December 2011)

On this occasion, the park service was denying the equivalence of valuation by focusing on savings but not costs. The state picked up on it. The official review of closure said that the park service's reasoning was opaque and that they were, in fact "insensitive to the public and private costs associated with the closure of particular park units, and inflexible in their consideration of cost saving alternatives to park closure" (Assembly Committee on Appropriations, 25 May 2012). Indeed, there were practical reasons why closing the park would be difficult, including the fact that a county road runs right through the site and the potential damage to the historical fabric of securing the buildings mechanically.

While a critique of the list-making process by legislators may seem like a step in the right direction, it nevertheless upholds the possibility of value's effective distribution. It maintains the wonderfully seductive notion of a perfect system where value can be properly detected, distributed, protected and enjoyed. And while that gives us comfort, it also serves to distract us from broader political pressures to push through a politically driven institutional overhaul.

List-making gave licence for lawmakers to reduce staffing and transform parks into what they called an "enterprise-based" organisations with a "more business-minded" and "entrepreneurial revenue enhancement strategy". These are the terms used by the Little Hoover Commission, which produced a report titled Recapturing Excellence in California's State Park System, which has since developed into a wholesale reassessment of the California State Park system, named Parks Forward. Parks Forward was an 18-month initiative with public forums and expert meetings geared to address the financial, operational and cultural challenges it faced by California State Parks to ensure the system's long-term viability. But for many, this approach runs counter to reasons why parks were established in the first place:

> "There's this greater purpose than profit, that's the whole point that they were set up in the first place. And parks were never meant to raise money, it was never intended that the parks would turn a profit, it's not until very recently that somebody decided that they should. People have a hard time defining why they need parks, they just know that when they want to get out of the city, they can look on a map and go, 'there's a green spot on the map, I wonder what's there, let's go there'".
>
> (Interview with supervising ranger, 18 June 2011)

But Little Hoover maintained a solution driven around the business case by advocating for a "new objective criteria to be developed through a public process for determining which sites truly hold state-wide significance" (Little Hoover, xi). While this, in theory, allowed for the possibility for state park expansion, the expectation was exactly the opposite.

Landrum (2005) notes in his study of entrepreneurialism in America's State Parks a long-running, often-class-based antagonism for over-commercialisation in public parks that he describes as an aesthetic consideration. More intriguing and pertinent to this current restructure is commercialisation that is driven for political reasons:

> Of particular concern is the tendency in some states to force privatisation on the state parks for what must be seen as primarily political reasons. Although such practices have caused serious problems at times, the state park administrators tend to be more pragmatic (and discreet) in citing their purposes for privatization. Three main reasons stand out: greater economy, greater efficiency, and necessity, in that a desired project or program could not have been undertaken otherwise.
>
> (Landrum, 2005, p. 30)

The Parks Forward review is less about site-specific case-by-case pragmatics of value and more about championing the merits of a major long-awaited recalibration that will more accurately reflect contemporary demands. And yet appeals for more accuracy in value's measurement distract us from appreciating it as relational, where every declaration of value requires its diminishment elsewhere. Understanding value as relational requires us to be far more suspicious about claims for its detection through ever-more-sophisticated mathematical models and to map out instead how value furthers the interests of those overseeing it.

Here we get to the nub of the issue. A belief in the existence of value as something external to our attempts to measure it can be mobilised to achieve all sorts of things. It can, especially, work to disguise value's equivalence and the necessary diminishment that comes along with value's declaration.

One ranger at Malakoff Diggins described the equivalence that gets missed in the new vision for state parks as a vibrant and sustainable system – at least while it remains open:

> It's not effective to take people on a nice walk in the woods if you don't tell them in nine months "you can't come here because this park will be closed" or if you don't tell them that "you can walk here, but in one mile you have to stop because it's going to be clear-cut for logging" which is going to be happening right over there in another year or so. So, I think people need to understand that even once you've protected something it's not protected forever and it's not safe from budgets cuts. It's not safe from greed; it's not safe from private interest.
>
> (Personal interview with regional level state park employee, 17 June 2011)

On 28 June 2012, just before the busy Fourth of July holiday week and two days before the planned deadline, California officials announced that 65 of the 70 state parks originally marked out for closure would remain open. The governor signed a bill allocating new funds. And the state reached a number of agreements with non-profit-making organisations and local governments to support the running of its parks. During a conference call with reporters, Ruth Coleman, director of the state parks system, said, "We have re-energized the people who love parks, and they are stepping up and contributing to parks in all sorts of ways". It was clear, though, that one of the major ways "the people" would be contributing was financially since attached to the package was US$13 million for schemes aimed at increasing revenue through fee machines that accept credit and debit cards and improvements to chargeable accommodation facilities such as cabins and campgrounds.

References

Assembly Committee on Appropriations, (2012) Committee Analysis of AB 1589, 25 May.

Baudrillard, J. and Petit, P. (1998) *Paroxysm: Interviews With Philippe Petit*. London: Verso, p. 4.

Bourdieu, P. (1986) 'The Forms of Capital', in Richardson, J.G. (ed) *Handbook of Theory and Research for the Sociology of Education*. New York: Greenwood Press, pp. 241–58.

California Department of Parks and Recreation (1967) *Master Plan Narrative, Malakoff Diggins State Historic Park*. Sacramento: Interpretive Services Section.

Castree, N. (2008) Neoliberalising nature: Processes, effects, and evaluations. *Environment and Planning A*, 40(1), pp. 153–173. https://doi.org/10.1068/a39100.

Clarke, D. (2009) 'The Limits to Value', in Smith, S., Pain, R., Marston, S. and Jones, J.P. (eds) *The SAGE Handbook of Social Geographies*, London: SAGE, pp. 253–268.

Cresswell, T. (2012) 'Value, Gleaning and the Archive at Maxwell Street, Chicago', *Transactions of the Institute of British Geographers*, 37(1), pp. 164–176.

Dartik, C. (2011) Comments by Executive Director of South Yuba River Citizens League in Town Hall meeting broadcast live from Miners' Foundry, Nevada City, by Radio Station KVMR on 12th December and transcribed by Gareth Hoskins.

Dewey, J. (1922) *Human Nature and Conduct: An Introduction to Social Psychology*. London: Carlton House.

Dewey, J. (1939) 'Theory of Valuation', in Neurath, O. (ed) *International Encyclopedia of Unified Science*, 2(4), Chicago: University of Chicago Press, pp. vii+67.

Egenhoff, E.L. (1965) *Scenic Resources of California: A Page from History*. Sacramento: Mineral Information Service.

Gold, J. and Burgess, J. (1982) *Valued Environments*. Boston: Allen & Unwin.

Harrison, R. (2013) 'Forgetting to Remember, Remembering to Forget: Late Modern Heritage Practices, Sustainability and the "Crisis" of Accumulation of the Past', *International Journal of Heritage Studies*, 19(6), pp. 579–595.

Harvey, D. (1982) *The Limits to Capital*. Oxford: Blackwell.

Hetherington, K. (2008) *Capitalism's Eye: Cultural Spaces of the Commodity*. New York: Routledge.

Hoskins, G. (2016) 'Vagaries of Value at California State Parks: Towards a Geographical Axiology', *Cultural Geographies*, 23(2), pp. 301–319.

Landrum, N. (2005) 'Entrepreneurism in America's State Parks', *The George Wright Forum*, 22(2), pp. 26–32.

Latour, B. (1993) *We Have Never Been Modern*. MA: Harvard University Press.

Law, J. (1993) *Organizing Modernity*. Oxford: Blackwell.

Lee, R. (2006) 'The Ordinary Economy: Tangled Up in Values and Geography', *Transactions of the Institute of British Geographers*, 31(4), pp. 413–432.

Miller, D. (2008) 'The Uses of Value', *Geoforum*, 39(3), pp. 1122–1132.

National Park Service (1972) Department of Interior National Register Inventory Nomination Form for Malakoff Diggins – North Bloomfield Historic District 30 Nov 1972. Available at National Archives Catalogue: https://catalog.archives.gov/id/123860133. (Accessed 10 October 2019).

Nietzsche, F. (1972) *The Antichrist*. New York: Arno Press.

Park Advocate, (2011) email correspondence with author 13 December.

Polanyi, K. (2001) *The Great Transformation: The Political and Economic Origins of Our Time*. Boston: Beacon Press.

Renda, M. (2011) *Questions Surround California Parks Closure List. The Union*, 3 November 2011. Available at: www.parkwatchreport.org/article.html?pub=news&query=&art=2647 (Accessed 2 July 2014).

Robertson, M. M., & Wainwright, J. D. (2013) 'The Value of Nature to the State', *Annals of the Association of American Geographers*, 103(4), pp. 890–905.

Smith, J. (2002) 'The Place of Value', in Agnew, J. and Smith, J. (eds) *American Space/ American Place: Geographies of the Contemporary United States*. Edinburgh: University of Edinburgh Press.

Smith, N. (1984) *Uneven Development: Nature, Capital, and the Production of Space*. Oxford: Blackwell.

Soja, E. (1980) 'The Socio-Spatial Dialectic', *Annals of the Association of American Geographers*, 70(2), pp. 207–225.

Sprinkle, J. H. (2014) *Crafting Preservation Criteria: The National Register of Historic Places and American Historic Preservation*. London: Routledge.

Sullivan, S. (2013) 'Banking Nature? The Spectacular Financialisation of Environmental Conservation', *Antipode*, 45(1), pp. 198–217.

Weiser, M. (2011) 'State's Park Closure Criteria Murky, Assembly Panel Told', *The Sacramento Bee*, 2 November. Available at: www.sacbee.com/2011/11/02/4023505/states-park-closure-criteria-murky.html (Accessed 2 July 2014).

White, R. (2000) 'The Problem With Purity', *Tanner Lectures on Human Values*, 21, pp. 211–228.

Yusoff, K. (2011) 'The Valuation of Nature: The Natural Choice White Paper', *Radical Philosophy*, 170(Nov/Dec), pp. 2–7.

Part III
Practising value

12 Unsettled value

Reidentifying tobacco industrial heritage in eastern Taiwan

Han-Hsui Chen

This chapter examines how heritage value is produced in Taiwan by focusing on the designation and preservation of tobacco agricultural systems. It is based on interviews and documentary research conducted with farmers, residents and policymakers in Fonglin township and Jian township, both historical tobacco settlements in eastern Taiwan. Having been introduced to Taiwan by the Japanese during their colonisation, tobacco agriculture had thrived for more than 70 years before it began to suffer a steady decline from the 1980s. Following the awareness of protecting local heritage, the residents of Fonglin and Jian looked to their once-thriving tobacco industry as a source of commemoration. Both Jian and Fonglin needed to carve out a unique heritage offer, distinctive from that of the main tobacco-growing region in western Taiwan, which had already been recognised as a post-productivist heritage opportunity. Thus, they chose to distinguish their townships by means of a broader historical context – interpreting their tobacco industry more directly to the Hakka industry and culture.

In this chapter, I argue that Taiwan's heritage is more complex and dynamic than people usually perceive, the interpretation of heritage is never simple and stable; in contrast, it is shaped by contemporary imperatives and ambitions. The initial action to protect heritage taken at the beginning of the post-war period affirmed Taiwan's inheritance of Chinese culture and cultivated an imagined Chinese identity. This led to the marginalisation of mundane, regionally focused, everyday heritage, such as local tobacco heritage.

Jian and Fonglin, both historical tobacco settlements in eastern Taiwan, illustrate how the value of local tobacco heritage was identified and interpreted by the local residents following the island's social transition from autocracy to democracy. Later, tobacco heritage's value is reshaped in relation to the practice of Hakka culture recultivation and preservation.

The chapter begins with a brief review of heritage literature to situate its theoretical context, before proceeding with a short history of Taiwan to illustrate its political and social transition. This is followed by the main analysis of the unsettled value of national and local heritage of Taiwan. The analysis focuses primarily on the national context of Taiwan's heritage sector and its relationship with the People's Republic of China. The effect of this complex relationship on the formation of Taiwan's national identity and ideology is examined, together with how it was connected to the emergence of a formal heritage policy in Taiwan, whereas local heritage was silenced against the backdrop of more direct targeting of priorities for national heritage. Second, the chapter reveals stories that illustrate how Taiwan's heritage policy was shaped by social and political change in relation to historical reinterpretations and local contributions to heritage construction. Finally, Jian's and Fonglin's respective tobacco heritages are taken as examples to demonstrate how contemporary local residents identify with their tobacco-related legacy and reinterpret their broad historical and cultural engagement to the Hakkanese identity in order to secure a "settled" value of their tobacco heritage.

Theoretical understandings of heritage

Heritage is a concept that has long been identified as valuable personal belongings inherited from past generations (Hardy, 1988; Lowenthal, 1996; Tunbridge and Ashworth, 1996), and this extends to the idea of public treasure belonging to the state or to the whole world. According to the definition given by UNESCO, "heritage is our legacy from the past, what we live with today, and what we pass on to future generations" (UNESCO, 2019). Geographers who study heritage have provided a more nuanced description of heritage, writing that "it is a view from the present, either backward to a past or forward to a future" (Graham et al., 2000, p. 2). These two statements imply that heritage is not fixed but instead constantly changing over time, becoming more and more complex as it engages with contemporary requirements of economic development, social inclusion and regional regeneration (Urry, 1994; Graham et al., 2005).

The critique of official politics and the exploration of the relationship between nation, national identity and heritage are part of a huge sub-field in heritage studies (Massey, 1995; Harvey, 2003; Graham et al., 2005; Smith, 2006). The politics of heritage can be approached from different perspectives, namely the politics of time, which can refer to the period in history chosen to commemorate; the politics of space, which alludes to the place nominated as heritage; the politics of stories, which is relevant to the narratives selected; and the politics of objects, related to the kind of material perceived to have historical and cultural significance. This is echoed by several studies, the best known of which is perhaps Hobsbawm and Ranger's edited volume *The Invention of Tradition* (1983). Hobsbawn and Ranger affirm that historical interpretation serves a particular political purpose, which is primarily to shape a specific national identity and to form what Anderson famously referred to as imagined communities (1983). Johnson (1996) in her key critical work on the contested historical representation of a big house in Ireland employs analytical public literature (the Big House novels) and historical tours of Strokestown Park House to illustrate that heritage is filled with plural meanings, since the past is selected and reinterpreted by people's understandings and approaches to their history. All the aforementioned studies demonstrate that the value of heritage is often unsettled and contingent because it is constructed and shaped by how people understand the past and interpret history.

Historical background of Taiwan

The history of Taiwan dates back around five hundred years, and according to archaeological excavations, its pre-historical period can be traced back more than 30,000 to 50,000 years, when Indigenous people first settled on the island. The initial wave of Han Chinese migrants in the 17th century consisted of two major groups, namely Min-nan and Hakka, and the second wave of Chinese migrants arrived after WWII, when the Kuomintang (KMT, or Chinese Nationalist Party) retreated from the mainland to Taiwan and brought their followers.

The Qing Dynasty on the mainland lost the First Sino-Japanese War and signed the Treaty of Shimonoseki in 1895, when Taiwan became colonised by the Japanese. This period of occupation ended in 1945, after the Second World War, when a governor was assigned to Taiwan by the KMT regime, which was based on the mainland. At that time, Chiang Kai-shek, the KMT leader, urged the people to fight the Communist Party for the possession of mainland China. However, as the Communist Party became stronger and stronger, Chiang Kai-shek and his people were exiled to Taiwan, which he conceived as being both a shelter and a military base to facilitate the KMT's return to the mainland. Chiang Kai-shek was a military general, and Taiwan became an autocratic state dominated by him. Chiang Kai-shek's KMT heavily promoted the connection between the mainland and Taiwan and advocated unification. Then, in 1986, a second political party, the Democratic Progressive Party (DPP), was formed by advocates of democracy and an independent Taiwan, according to the interpretation of the Cairo Declaration, the Potsdam Declaration and the Treaty of San Francisco (the Treaty of Peace with Japan). The argument for an independent Taiwan or unification (Taiwan and the mainland) was more than a domestic issue; it was entangled with a controversy that began in 1945 about the political relationship between Taiwan and the People's Republic of China. Although Taiwanese society has now transitioned from autocracy to democracy, where state administration is transferred between different political parties through public elections, the political relationship with the People's Republic of China is still a major issue today, which heavily influences the construction of Taiwan's national identity and heritage policy.

Constructing national identity and the emergence of heritage in Taiwan

The KMT spared no effort in connecting Taiwan with mainland China by constructing a cultural consciousness in an attempt to legitimise its occupation of the island, and one way of doing this was to destroy all traces of the Japanese colonial legacy. Other methods included claiming that all Chinese treasures had been moved to Taiwan from the mainland and stored at the National Palace Museum. In this context, the National Palace Museum represented not only the Chinese culture but also represented that the KMT regime had inherited Chinese traditions and was the legitimate ruler of Taiwan, and it reinforced the Chinese national identity (Chun, 1998). This could be interpreted as being the start of top-down heritage protection in Taiwan. The National Palace Museum still has a strong reputation and attracts a large number of tourists every year, since grand and elite heritage accredited as "national" tends to be more attractive to visitors than are local heritage sites, like the old tobacco barns in tobacco settlements or Indigenous residences.

Furthermore, the inheritance of Chinese traditional culture was regarded as a tourist resource, and the KMT authorities began historical preservation work in an effort to attract more international tourists. However, the architectural style of the historical preservation work was altered in some cases to imply that the history of Taiwan was embedded in the Chinese cultural context. This was a controversial issue between heritage protection and national politics: preserving the existing history while recreating new historical value. Sidorov's work on examining the politics of scale in the reconstruction of the Cathedral of Christ the Saviour in Moscow provides an insight into how the architectural scale related to the construction of this national monument was woven into the transition of the state and society (Sidorov, 2000). Analogous to the case of Moscow, the architectural style of the old Taipei city gates was reconstructed by the authorities to emphasise the connection between architecture and national imagination.

The old Taipei city was built in 1879 by the Qing dynasty, and each city gate had a traditional south style tower built over it. Most of the city wall was demolished by the Japanese authorities in their urban regeneration plan, and only four city gates remained.[1] However, the original architectural style of the city gate towers was reconstructed from the traditional south style to the north palace style in the 1966 historical conservation work in order to demonstrate Chinese culture (Kuo, 2009; Yen, 2006). The traditional south style of architecture was seen as a kind of vernacular architecture, which was insufficiently powerful to symbolise Chinese national identity. The north palace style, in contrast, could symbolise it, because the capital of the Chinese empire was always located in the north. The transition of the architectural style of the old Taipei city gates illustrates that the production and reading of heritage-scapes "are political in the proudest sense of the term, for they are inextricably bound to the material interests of various classes and positions of power within a society" (Duncan, 1990, p. 182). The KMT regarded heritage as a tool to shape the national identity and affirmed the inheritance of Chinese culture in post-war Taiwan, on the national and the international scales.

To secure the regime, the KMT applied autocratic domination in Taiwan; however, there were slight changes in society when the introduction of various Western thoughts, including the concepts of modernity, post-colonialism and citizenship, by people who had studied or lived abroad invoked a variety of critical reflections. Architectural scholars advocated the concept of the vernacular, claiming that there should be more focus on an ordinary cultural and locally built environment rather than only promoting Western modernity and traditional Chinese culture in the architectural realm. Autocratic domination was gradually relieved by the advocacy of democracy and liberty, and heritage protection emerged from this fragile democracy.

In 1976, a movement to protect Lin An-Tai Old House was the first in the fight to retain a historical building against the urban plan in post-war Taiwan. The movement was unsuccessful, and the old house was demolished and reconstructed in another place a few years later. However, it had far-reaching effects in raising awareness of historical conservation and was regarded as a turning point that led the Executive Yuan to publish the Cultural Heritage Preservation Act in 1982 (Yen, 2005). Although this led to a new age and gave the work of heritage preservation a regulation with which to comply, the designation of cultural heritage under the Cultural Heritage Preservation Act was still greatly constrained by political concerns. Before 1985, the designation of historical sites was limited to temples, old city walls, mausoleums and habitations, the historical context of which all related to mainland China. The authorities designated these historical sites as part of the national heritage, to remind people that Taiwan had an intimate and unforgettable connection with the mainland. As Yen points out,

> National heritage is a tool to produce and reproduce the image of national history and geography. It is also a symbol of national awareness. . . . to evoke the tradition of the past to identify the geographical boundary of modern nations and their citizens' consciousness of identity.
>
> (2005, p. 7, author's translation)

In contrast, historical sites that could not be interpreted as part of the history of the great Chinese empire, such as the legacy of the Indigenous peoples and the Japanese colonisation, were ignored and omitted from the list of national heritage sites. Also, mundane, regionally focused, everyday heritage, such as local tobacco and sugar industrial heritage, were marginalised.

Local heritage and community empowerment

Another turning point that led to a dramatic increase in the awareness of recognising and protecting local heritage was when Lee Teng-hui, a Taiwan-born KMT politician, inherited the position of president in 1988. Lee promoted the movement of "Taiwanisation", which echoed the advocacy of Taiwanese consciousness, in an attempt to reconstruct Taiwan's national identity to combat the country's diplomatic isolation.

Community empowerment was another main discourse that emerged during Lee's presidency. It was proposed by anthropologist Chen Chi-Nan when he was vice-minister of the Council for Cultural Affairs[2] between 1993 and 1997. Community empowerment played a significant role in social transition, cultural protection and environmental sustainability in Taiwan by advocating the concept of symbiosis, the strategic vision of the country under Lee's presidency. Community empowerment was crucial in protecting local heritage, since most community empowerment endeavours were triggered by the recovering of local history and heritage and the merger of local identity and cultural awareness (Zeng, 2007). Heritage expert Mr Lin echoes community empowerment expert Mr Zeng's perspective of the contribution of community empowerment to heritage practice. According to Lin, community empowerment was usually enacted in three ways, the first of which was cultural and economic regeneration led by founding cultural spaces; the second was culture-led local revitalisation by establishing local museums; and the third was rediscovering local culture by means of a cultural and historical society organised by local residents. Here, heritage practice is deemed to enhance the awareness of heritage protection or improve the grassroots involvement in heritage works (Lin, 2011). This policy still has a huge influence on current society and is the basis of cooperation between different agencies in the central government.

There appears to have been several facets of this policy, the first of which is that it was the authorities' lead policy, which was initially promoted at national scale. It was used to shape a new national ideology in response to political concerns. Second, from the perspective of the promoter, it was designed to improve an awareness of citizenship. Third, community empowerment was conceived as a locally focused policy, since it aimed to stimulate locally motivated action to improve the local cultural environment; therefore, the idea of community empowerment became the main type of both social and cultural policies, encouraged by different authoritative departments by providing subsidies to support community empowerment projects. It was in this context that Jian township's and Fonglin township's respective tobacco-related historical regeneration projects emerged.

Valueing Jian's tobacco heritage

Jian township was the first Japanese immigrant village established by the Japanese authorities in 1910 during the colonial period. Tobacco was introduced by the colonists to immigrant villages as a new economic crop. The Japanese authorities monopolised the production of tobacco in 1914, and it became an important source of state revenue. In 1918, Jian tobacco trade centre was built for the monopoly bureau and tobacco farmer to trade tobacco leaves. Because of this, the tobacco industry thrived, and the original trade centre was rebuilt and extended in 1961. Following the industry decline, the trade centre went out of use in 1992.

Jian township was the first place that Japanese immigrants cultivated tobacco in Taiwan, so the historical and cultural significances are quite strong in terms of the commemorating the development of tobacco cultivation in Taiwan. The value of tobacco legacy was first recognised by Nanhau, a community development association in Jian township, rather than by Jian Township Office. The first conservation work on tobacco agricultural heritage was family Siao's tobacco barn, which was one of the four tobacco barns still remaining in the local area when tobacco industry ended. Siao's tobacco barn was conserved by the Nanhau community development association in 2005. The association setup a working holiday project and cooperated with Taiwan Environment Information Centre and Bayer Taiwan Ltd, which attracted 30 people come to join this tobacco barn conservation working holiday. For Nanhau community development association, the protection of tobacco heritage was for more than just preserving local tobacco history; it was also believed that it could help identify local characteristics and improve local tourism development. However, because the idea of tobacco barn's value was dissimilar between local people and local township office, the township office decided to end the funding of tobacco barn maintenance expenses in 2012. Nowadays, visitors can only stay outside of the conserved tobacco barn: they are not able to enter to see the interior of the tobacco leaves curing space. On might conclude that the Jian Township Office did not care about local tobacco heritage. Nevertheless, that same year, Jian Township Office set up a project to apply for funding from the Hakka Affairs Council to preserve another local tobacco barn, Jian tobacco trade centre, the first tobacco trade centre built in Taiwan.

The Hakka Affairs Council was formed in 2001, following the development of democracy in Taiwan, when cultural diversity became respected at both national and local levels. The primary mission of the Hakka Affairs Council is to revitalise the Hakkanese language and culture and build the Hakkanese identity. Jian Township Office claimed that Jian is a Hakka town. Jian has the largest Hakka population (over thirty thousand) among all the townships located in Hualien Country (Chen, 2006).[3] As mentioned earlier, the society of Taiwan has had a dramatic transition: since the beginning of the post-war period, the Minnan and Hakka people were marginalised from the mainlanders who retreated with KMT regime. Following the social transition, different groups of people gradually gained more and more respect; at the same time, in line with the community empowerment policy, local authorities and residents began to recultivate historical and cultural roots and, further, reshape the township as a Hakkanese settlement.

To get funding from the Council for Hakka Affairs, the project applicant must be related to Hakkanese culture. Therefore, the tobacco industry was interpreted as a Hakkanese industry and Jian Township Office included the tobacco trade centre conservation project into a huge township regeneration plan titled the Jian Township Jhugaocuo Hakka Settlement Cultural Industry Centre Development Plan to apply for funding. The tobacco trade centre will be reused as a cultural creativity education centre after its conservation. According to the introduction of the Jian Township Jhugaocuo Hakka Settlement Cultural Industry Centre Development Plan, the slogan is "Hakka Settlement Building, LOHAS Jian".

> The mission is to establish Hakka industrial exchange and cooperation centre, and set up tourist industrial service system that based on the action of Hakkanese characteristic cultivation, in order to trigger industrial development in Jian. There are two parts of the plan, one is the hardware construction, while the other is software management. In terms of the hardware construction, the social benefit hall in Jian park will be conserved as a base for output service, such as developing industrial exchange and cooperation; while Jian tobacco trade centre will be identified as an input centre for cultural creative educational centre.
>
> (Jian Township Office, 2012)[4]

The introduction of the project clearly shows that the conservation of tobacco trade centre was conceived as a cultural element to support the shaping a new

Hakka settlement: Lohas Jian. In this context, the value of tobacco heritage is degraded in terms of highlighting the Hakka characteristics.

From the local scale, the private tobacco heritage (Siao's tobacco barn) was regarded as heritage that has some cultural and tourist value to be protected; however, when it competed with the public tobacco heritage (the tobacco trade centre), the public won the attention from the local authorities. From a national scale, only highlighting the tobacco historical characteristics seems not enough to win public funding for conservation and reuse. The township office included local tobacco heritage into the Hakka cultural development project to win the funding from the Council for Hakka Affairs. However, the value of tobacco heritage changed from commemorating once-thriving local tobacco agriculture to celebrating Hakkanese industry and fostering local cultural creativity.

Valueing Fonglin's tobacco heritage

Fonglin's tobacco history began in 1918, when it was the third village in Taiwan inhabited by Japanese immigrants during the Japanese colonial period. The cultural and historical significance of tobacco agriculture has been recognised as heritage by the people of Fonglin since 2002; however, the understanding of the value of tobacco heritage varies, and this causes tension between different ways of interpreting and representing the local agricultural history.

Fonglin Township's first historical regeneration plan, which was formed in 2002, was subsidised by the Construction and Planning Agency of the Ministry of the Interior, under the Townscape Renaissance Project, a competitive funding opportunity. Fonglin used the acquired funding to conserve two abandoned tobacco barns, which had originally been part of the town's historical regeneration plan. However, the tobacco barns project was now controlled by the central government, and the conservation work had to be completed in line with the opinions of the committee members of the Construction and Planning Agency of the Ministry of the Interior. After the conservation, one of the tobacco barns became famous for its Western appearance, a Tudor-style architecture with an outdoor seat in the style of the Spanish architect Gaudi. This kind of historical reconstruction, which was led by the examining committee, made many people suspicious.

According to a Fonglin township official, one of the committee members responded to local suspicion by asking why tobacco barns should look like a "traditional tobacco barn" (Mr Liu, interviewed by the author, 13 April 2010). This conservation led to a heated debate, since Western-style architecture was alien to the local cultural context. For example, a local tobacco farmer claimed, "The Western-style conserved one is not a tobacco barn. Its shape is not the same as the original" (Mr Huang, interviewed by the author, 25 May 2011). The tobacco farmer thought that the tobacco barns should have been preserved according to their original style, which they perceived to be the value of conserving the tobacco heritage. Giving an old building a Western appearance, such as the Tudor-style tobacco barn, was inharmonious and did not fit the local cultural context. The tobacco farmer even claimed that it was not a tobacco barn. The case of this alien tobacco barn illustrates the tension between different values of heritage (traditional versus creative) from the local to the national scales (local people versus the committee) that made the meaning of tobacco heritage in Fonglin complex and controversial.

After two tobacco barns had been conserved by the Hualien County government, funded by the Construction and Planning Agency of the Ministry of the Interior, the Fonglin Township Office decided to conserve the other two tobacco barns themselves. In 2006, Fonglin was chosen as one of the 15 Hakka settlements for preservation by the Council for Hakka Affairs,[5] and the township office acquired a grant from the council to conserve the tobacco barns led by the township office. They tried to restore the tobacco barns by applying traditional techniques and materials and keeping as close as possible to the original format. Most of the projects related to Fonglin's tobacco heritage preservation are funded by the Hakka Affairs Council. As one of Fonglin's officials said in an interview,

> Because most of the population here is Hakkanese, we applied for funding from the Hakka Affairs Council rather than Council for Cultural Affairs because we have no cultural historical sites. To be honest, the tobacco barns do not belong to the Hakka, just because most Fonglin people are Hakkanese. Many of them cultivated tobacco, so they built tobacco barns, . . . that is why we regard tobacco as one of the elements of the Hakka and a characteristic Hakka industry.
>
> (Fonglin Township Office official Mr Wong, interviewed by the author, 3 March 2011)

This comment illustrates that the legacy of the tobacco industry is not officially conceived as heritage, which meant that Fonglin was unable to obtain any finance from the Council of Cultural Affairs. Therefore, the township office attempted to find a way forward by interpreting the tobacco heritage as an element of the Hakka and thereby successfully obtained several subsidies from the Hakka Affairs Council to protect the local heritage.

Fonglin's heritage practice of tobacco agriculture illustrates the unsettled value of heritage. The cultural value of the tobacco-related legacy was not identified until the increase in the awareness of protecting local heritage following the policy of community empowerment. Fonglin has recultivated its Hakka town image and reinterpreted the close relationship between Hakkanese culture and tobacco history.

From a national perspective, Fonglin's tobacco heritage is marginalised, since tobacco barns are not acknowledged as being *national* heritage by the authorities, and they cannot obtain financial resources from official heritage preservation institutions like Council for Cultural Affairs, even though the local people firmly believe that the tobacco legacy is their heritage. However, by cultivating their Hakka character, Fonglin was able to obtain a subsidy from Hakka Affairs Council to preserve both its local tobacco industry legacies. From a local perspective, Fonglin successfully reshaped its local identity as a Hakka settlement, and the tobacco industry is conceived as Hakkanese agriculture.

Conclusion

This chapter examined how heritage is valued and interpreted by different groups at different times to serve different political, social and economic purposes. Action to protect heritage in Taiwan was initially carried out on the basis of the state's political concern to affirm the inheritance of Chinese culture and to shape an imagined Chinese identity in Taiwan at the beginning of the post-war period. In this context, local heritage practices, such as Jian's and Fonglin's respective tobacco commemorations, were alienated from the broader context of national heritage, and at the same time, the KMT regime spared no effort in destroying the Japanese legacy and the island's native culture. The concept of heritage practice emerged following the transition of Taiwan's social and political environment, and heritage interpretation became a contest between those who advocated a Chinese-centred ideology and those who promoted a Taiwan-centred one. The consciousness of the need to protect heritage became stronger after the policy of community empowerment, which attempted to reform people's sense of being Taiwanese and their cultural identity.

Heritage practice at the local scale seems to have been absent until the policy of community empowerment was introduced. This supports Kevin Lynch's perspective of preserving "great history" versus "micro history", when he says,

> Most historical preservation, focused as it is on the classic past, moves people only momentarily, at a point remote from their vital concerns. It is impersonal as well as ancient. Near continuity is emotionally more important than remote time, although the distant past may seem nobler, more mysterious or intriguing to us. There is a spatial simile: feeling locally connected where we customarily range is more important than our position at a national scale, although occasional realisation of the latter can impart a brief thrill. In this sense, we should seek to preserve the near and middle past, the past with which we have real ties.
>
> (Lynch, 1972, p. 61)

From Lynch's perspective, historical preservation should begin with a story that has an intimate relationship with ordinary people, and this kind of historical preservation "strengthens our own sense of identity" (Lynch, 1972, p. 61). Atkinson adopts "the democratisation of memory" to echo this standpoint, and he further explains that such democratisation involves steering "attention from high-profile heritage sites towards the less spectacular, quotidian and mundane places where social memory is produced and mobilised" (2008, p. 382). In this vein, community empowerment provides a means for "communities to reflect on their individual and collective past" (ibid., p. 383) and to show their value of local heritage.

At the beginning of the post-war period, people were, understandably, not interested in historical preservation, since those works were focused on the inheritance of Chinese culture rather than on telling the story of people's ordinary lives. This led to a huge gap between the authorities and ordinary people in their respective understandings of heritage. Historical preservation was overwhelmingly seen from a political perspective rather than a social and cultural one, which constrained the awareness of the protection of local heritage and local identity.

In the case of Jian and Fonglin, they connected their historical tobacco industry with Hakka settlements in order to claim a local interpretation of their heritage and to win financial support to complete the conservation work. Tobacco heritage in these two townships are given a new value by revisiting the local history and engaging the Taiwan's social transition. However, the more Hakka culture is highlighted, the more degraded the tobacco's historical and cultural significance. These all complicate the view of heritage in Taiwan and reveal an unsettled value of heritage.

Notes

1 Although the city wall was destroyed, the four city gates remained and were designated, by the Japanese authorities in 1935, as historical legacies.
2 The Council for Cultural Affairs was established in 1981 and was reorganised as the Ministry of Culture in 2012.
3 The Hakkanese people are a group of Han Chinese, who came to Taiwan from the mainland during the 17th century.
4 The official website of Jian Township Office: www.ji-an.gov.tw/charmjian/hakka-themed-infrastructure/39330. [Accessed 18 January 2019].
5 Fonglin is conceived as a Hakka settlement, because more than 80% of its population is Hakkanese.

References

Anderson, B. (1983) *Imagined Communities: Reflections on the Origin and Spread of Nationalism*, London: Verso.
Atkinson, D. (2008) 'The Heritage of Mundane Places', in Graham, B. J. and Howard, P. (eds) *The Ashgate Research Companion to Heritage and Identity*. Aldershot: Ashgate, pp. 381–395.
Chen (2006) *The Imagination and Practice- The Hakka Culture Representation of Jian Township*. Unpublished masters thesis, Department of Ethnic Relations and Culture, National Dong Hwa University.
Chun, A.J. (1998) 'The Cultural Industry as National Enterprise: The Politics of Heritage in Contemporary Taiwan', in Dominguez, V. and David., W. (eds) *From Beijing to Port Moresby: The Politics of National Identity in Cultural Policies*, Taylor & Francis Group, pp. 77–114.
Duncan, J.S. (1990) *The City as Text: The Politics of Landscape Interpretation in the Kandyan Kingdom*. Cambridge: Cambridge University Press.
Graham, B.J., Ashworth, G.J. and Tunbridge, J.E. (2000) *A Geography of Heritage: Power, Culture, and Economy*. London: Arnold.
Graham, B.J., Ashworth, G.J. and Tunbridge, J.E. (2005) 'The Uses and Abuses of Heritage', in Corsane, G. (ed) *Heritage, Museums and Galleries: An Introductory Reader*. London: Routledge, pp. 26–37.
Hardy, D. (1988) 'Historical Geography and Heritage Studies', *Area*, 20(4), pp. 333–338.
Harvey, D.C. (2003) 'National' Identities and the Politics of Ancient Heritage: Continuity and Change at Ancient Monuments in Britain and Ireland, C.1675–1850', *Transactions of the Institute of British Geographers*, 28(4), pp. 473–487.

Hobsbawm, E.J. and Ranger, T.O. (1983) *The Invention of Tradition*. Cambridge: Cambridge University Press.

Johnson, N.C. (1996) 'Where Geography and History Meet: Heritage Tourism and the Big House in Ireland', *Annals of the Association of American Geographers*, 86(3), pp. 551–566.

Kuo, C.L. (2009) 'Conservation of City Architecture and Public Sphere in the Post-war Taiwan', *Journal of Architecture*, 67, pp. 81–96.

Lin, H.C. (2011) *Outline of Taiwan Cultural Heritage Preservation History in Taiwan*. Taipei: Yuan-Liou Publishing.

Lowenthal, D. (1996) *The Heritage Crusade and the Spoils of History*, London: Viking.

Lynch, K. (1972) *What Time Is This Place?* Cambridge, MA, London: MIT Press.

Massey, D. (1995) 'Places and Their Pasts', *History Workshop Journal*, 39(1), pp. 182–192.

Sidorov, D. (2000) 'National Monumentalization and the Politics of Scale: The Resurrections of the Cathedral of Christ the Savior in Moscow', *Annals of the Association of American Geographers*, 90(3), pp. 548–572.

Smith, L. (2006) *Uses of Heritage*. London: Routledge.

Tunbridge, J.E. and Ashworth, G.J. (1996) *Dissonant Heritage: The Management of the Past as a Resource in Conflict*. Chicester: Wiley.

UNESCO (2019) *UNESCO World Heritage Centre*, Available at: http://whc.unesco.org/en/about/ (Accessed 15 January 2019).

Urry, J. (1994) *Consuming Places*. London, New York: Routledge.

Yen, L.Y. (2005) 'Cultural Heritage in the Age of Globalisation: A Critical Review of Historic Preservation Theories', *Journal of Geographical Science*, 42, pp. 1–24.

Yen, L.Y. (2006) 'Time-Space Imagination of National Identity: The Formation and Transformation of the Conceptions of Historic Preservation in Taiwan'. *Journal of Planning*, 33, pp. 91–106.

Zeng, S.J. (2007) *The Community Empowerment in Taiwan*. Taipei: Walkers Cultural Print.

13 Locating value(s) in political ecologies of knowledge

The East Svalbard management plan

Samantha Saville

> While we stopped to survey lichens, we spotted a gosling amongst the rocks. Our
> excitement watching this cute, fluffy, photogenic chick turned to concern when it
> began to follow us. Were we disturbing nature? Were we leading it too far away
> from its mother? We decided to move on and "lose" our intrepid follower. A few
> minutes later we saw a polar fox carrying a gosling away. We were left sobered and
> full of questions we could not answer. "Was it our fault?" Did our presence make
> a difference? "That's nature for you, harsh". "At least Mr Fox got some lunch".
>
> (Field notes, 4 July 2013)

While, academically, we may have resolved the problem of our place in "nature-
culture" (Davis, 2007), in practice, as this extract illustrates, tensions and questions still arise, particularly in regions that are more "wild". In this chapter,
I explore such tensions through the example of wilderness management in Svalbard. This place says "welcome to the Arctic – as long as you leave no signs of
having been here!" (Governor of Svalbard, 2010, p. 8). I demonstrate that tracing
value processes – concentrating on what value does, the "work" it performs, how
it is practised and what relations are caught up in productions and circulations of
value – can be an illuminating analytical tactic capable of tracing links between
"everyday" micro-value practices and wider geopolitical and international environmental concerns. I argue that wilderness management in Svalbard can be
understood as part of a wider framework of value circulations and constructions.
These value practices are not yet fully incorporated into plans for protected area
management. I take knowledge production[1] in Svalbard as one part of an ecology
that also encompasses state regulations, tourists, residents and more than human
natures.

Throughout the chapter, I draw on extensive interviews and ethnographic experiences from research undertaken in the summers of 2013 and 2014. I begin by
introducing the guiding premises of working with value and values in this way.
Svalbard is then located, geographically, as a valued site of knowledge production. This is developed into an understanding of Svalbard as a place that acts as a
hub in an expanding network of value through the practice of knowledge production. Svalbard as a central point where many actors meet also exerts its own force

on those people producing knowledge of it and with it. Finally, I examine the case of the East Svalbard Management Plan as an example of tensions between different ways of enacting value and values. Here knowledge about Svalbard is used to support opposing strategies of wilderness management, and how knowledge is enrolled in ecological management becomes a key point of tension. A split between humanist and eco-centric ideological stances is apparent. However, an expanded examination of value circulations presented here further highlights the need to take value seriously in decision-making and as an analytical tool.

Locating a value approach

> Value emerges contextually and relates to the interests of those doing the valuing.
> (Bourdieu, cited by Cresswell, 2012, p. 167)

George Henderson (2013) notes, in his detailed evaluation of the concept of value in the writings of Marx's, the importance of paying attention to the work that value does and how this work in turn affects value. Henderson considers value to be performative and relational – a concept that can travel beyond the realms of labour and capital relations (2004). Similarly, anthropologists David Graeber (2001) and Daniel Miller (2008) and sociologist Beverley Skeggs (2014) are keen to focus not on what value is exactly but on what value *does*, how it is practised and what relations are caught up in the production of value and values. My examinations of value in Svalbard have been inspired by a combination of these ideas and the beginnings of a geographical engagement with value as being contingent, thoroughly peopled, political and constructed through the values of those caught up in the processes of valuation (Lee, 2006; Cresswell, 2012; Robertson and Wainwright, 2013).

Miller (2008) looks into the link between *value* as a quantitative, calculative term and *values* as qualitative and intangible. He argues that the two are not easily separated but are intimately related on a moving scale or bridge, and "this bridge lies at the core of what could be called the everyday cosmologies by which people, and indeed companies and governments, live" (Miller, 2008, p. 1,123). In his analysis of shareholder value and "best-value" policies in local governance, Miller (ibid.) shows how calculative value dominates yet prevents the ultimate realisation of societal benefits, due to a failure to accommodate qualitative values. Drawing on Miller's approach and the insights mentioned earlier, I seek to explore how value and values of knowledge production are practised in Svalbard and what this means in wilderness management. Methodologically, this has meant a focus on value from the start, with lines of investigation, questions and participant selection all geared towards exploring Svalbard through this lens.

Svalbard is an interesting case from the perspective of knowledge networks and circulation, given its specifically international environment and a recent shift towards becoming a "centre of calculation" and knowledge creation (Jöns, 2015). Paying attention to how value flows and how values connect between different stakeholders, on different levels through both "everyday" happenings and particular moments of decision-making, offers one way to address seemingly paradoxical impasses.

Svalbard as a valued site for knowledge production

Svalbard is an archipelago in the Arctic Circle between 74 and 81 degrees north, governed by Norway under a special international treaty since 1925. Unlike many Arctic areas of human settlement, Svalbard's population is non-Indigenous, cosmopolitan and transient. The majority of the 2,500 residents live in the Norwegian capital of Longyearbyen, and the active Russian coal mining town of Barentsburg houses four hundred workers, including scientific bases.

Svalbard's physical features of relative accessibility, arctic ecology and geological and glacial formations have attracted scientific expeditions from many nations. Svalbard has a history of exploration dating back to the 16th century. The extraction of natural resources, with the accompanying scientific, specifically geological, knowledge took precedence from the year 1600 to the modern coal mining of today (Avango et al., 2011). Non-mining-related scientific activity and tourism were marginal but ongoing until recently. Since the 1990s, Norway has actively sought to support, encourage and promote Norwegian science and international collaborations in Svalbard as part of its economic diversification away from a sole reliance on coal mining and active demonstration of sovereignty in Svalbard.[2] In 1993 the University Centre in Svalbard, UNIS, opened in Longyearbyen.

Svalbard now attracts a wide range of natural and physical scientists, having developed facilities in Longyearbyen and research bases in the former mining base of Ny Alesund, which hosts research stations for 13 nations. For climate scientists, its geographical position gives access to Arctic conditions in the east and the effects of the Gulf Stream in the west through the West Spitsbergen Current, as well as the Arctic Ocean and Greenland Sea – interesting too for marine biologists. Almost two-thirds of the nearly 62,000 km² land mass is glaciated, so glaciologists have a range of glacial processes, including surging glaciers, to attract them. Physicists can make use of the high latitudes and highly specified arrays to observe auroral conditions and other upper-atmospheric phenomena. Terrestrial biologists are occupied with rare and unique species such as the polar bear, Svalbard reindeer and myriad invertebrates, as well as the plethora of migratory birds that arrive in the summer. Arctic technology is a growing department at UNIS as well, combining the challenging climatic and physical conditions of the arctic with access to existing and developing industrial activities.

The research and education sector in Svalbard continues to grow. UNIS reports continued growth, with nearly eight hundred students attending and over one hundred staff members or adjuncts producing 153 published articles in 2017 (UNIS, 2013). The international research base in Ny Alesund also reports increased activity (Kings Bay AS, 2013). Operations out of Russian scientific bases in Barentsburg and Pyramiden are likewise rising and are seen as a key area of growth for the Russian presence in Svalbard. For the purposes of this chapter, I am focusing on the main settlement and access point of Longyearbyen, where UNIS, the Norwegian Polar Institute and associated research facilities are located: around one-third of the scientific activity takes place in this area.

A valued political site

> In Svalbard, scientific research must be regarded at least in part as informal diplomatic activity engaged in by Norway (which makes such research possible) and by other states (which fund researchers).
>
> (Grydehøj, 2013, p. 53)

Political analysts and both residents and visitors interested in Svalbard politics do not regard Norway's sovereignty of Svalbard as solid, with the Svalbard Treaty limiting the extent of power that Norway can have and leaving much open to interpretation (e.g. see Pederson, 2009; Grydehøj, 2013). Previous research has asserted that the expansion of tourism and research are a continuation of contests over sovereignty cloaked in discourses of energy security, climate change and local economic development (Timothy, 2010; Avango et al., 2011; Grydehøj, 2013).

In Norwegian policy documents relating to the Arctic, the political role of research is seemingly transparent: Norway's High North Strategy focuses on "knowledge, activity and presence" (Ministry of Foreign Affairs, 2007, p. 6) as key words, with knowledge development and knowledge partnerships being two of the five areas that are seen as important to "value creation" in the region (Ministry of Foreign Affairs, 2014). More specifically, research, knowledge and higher education were key focus areas for development in Norway's governance of Svalbard. Svalbard was "of vital importance as a platform for Norwegian and international research. . . . Although Svalbard must remain an attractive venue for scientists from around the world, Norway is to have a leading role and be a key player in the area of developing knowledge in and around Svalbard" (Norwegian Ministry of Justice and the Police, 2010, p. 11). The most recent White Paper for Norwegian policy in Svalbard recognises many of its previous development objectives have been met, while it remains an ongoing priority area (Norwegian Ministry of Justice and Public Security 2016).

Following Jöns (2015) in a brief Latourian analysis of this situation, in terms of knowledge circulation and networks, the importance of Svalbard to Norway becomes clear. Creating knowledge with/about Svalbard not only means that Norwegian scientists working there can be part of global knowledge networks and strengthen both Svalbard research centres and Norwegian mainland ones but also means that international scientists will find themselves encouraged to collaborate with Norwegian institutions through funding and logistical advantages. Therefore, more and more links and stronger relations with researchers and organisations in Svalbard and Norway are developed internationally. This, as Grydehøj notes, is one way international relations can play out through science, with science "naturalising" competition to some extent (Katz and Kirby, 1991). The importance of asserting sovereignty and the geopolitical significance of their/our activities is not lost on those in research institutions, from high-level managers to students and indeed beyond, as one member of the scientific community expressed:

> "Everything we do, everything everyone does on Svalbard, including the jan-itors is part of a geopolitical framework. Having contact with all these nations means that every word I say has to be weighed. . . . It influences everything that happens in Svalbard, including your work".
>
> (Interview, 27 May 2014)

Moreover, extra care appears to be taken that the practising of scientific activity here is made visible through material inscriptions such as signage, flags and branded equipment (e.g. see Figure 13.1).

Connected to the political need for a presence in and the prestige of being a leading research nation of the Arctic is environmental protection. Under the Svalbard Treaty, Norway is able to legislate for environmental preservation. Norway has certainly taken this opportunity and has high aspirations of Svalbard being recognised as one of the best-managed wilderness areas in the world and has applied for World Heritage Status. Over 65% of the landmass is part of a protected area through National Parks, Nature Reserves or Bird Reserves. Scientific knowledge has been key to establishing the significant legislation and protection areas throughout the archipelago, beginning in the 1970s, and remains an important tool in their ongoing management. However, by the penultimate White Paper, the environmental impacts of scientific activities were being weighed against the value of the knowledge they can produce.

> Research that is conducted ought to be of such a nature that it only or best can be conducted in Svalbard, and it must always take the vulnerability of the environment into consideration. This caution must go hand in hand with the acknowledgement that knowledge through research is necessary in order to achieve a reliable management of the natural wilderness in Svalbard.
>
> (Norwegian Ministry of Justice and the Police, 2010, p. 75)

Whereas the subsequent White Paper offers more nuance, at this time, environmental protection seemed the policy priority: "environmental considerations are to take precedence over other interests whenever they conflict" (ibid., p. 10).

Svalbard is clearly a highly valued site geopolitically, for Norway and other nations: not only for the opportunity to increase knowledge about the arctic but also for the political prestige or symbolic value of being an active arctic nation with high scientific and environmental standards. Behind this political and policy support is of course a large economic framework for both the scientific activity and the governance structures of environmental protection. In both cases, employment positions are considered very prestigious, have good salaries and other valuable benefits like subsidised accommodation. Academic positions are constructed to

Figure 13.1 The University of South Bohemia's research container, complete with signage and Czech, Russian and Norwegian flags and a Swedish Polar Research Secretariat jacket in the background

attract high-calibre staff, so employment at UNIS comes with a far larger research budget than colleagues on mainland Norway. Yet the economics are not, I argue, the driving force or at least not the most interesting aspects of this ecology.

A hub of opportunities: practice and values

The goal of Svalbard becoming an arctic hub is certainly aspired to by policymakers and some members of the Longyearbyen community. Honing in on knowledge production, it would seem this has already been achieved to some extent. The 2008–2009 Norwegian White Paper stated that "Svalbard has become a meeting place for the government's international network . . . Svalbard has become a land of opportunity for the development of knowledge" (2010, pp. 74–76). The most recent paper affirms this position. The opportunities available in Svalbard for learning, research, impact and access to field data, equipment and expertise were widely recurring themes in the interviews and conversations I had in and around the scientific community in Longyearbyen:

> "The list of politicians, business men, religious leaders and others that come here is very long. And we get to meet them and explain what we see and understand. So it's an opportunity to influence directly those who would listen that is not as readily available elsewhere".
>
> (Interview, 27 May 2014)

> "The only reason I keep coming back is because the geology is so good, the research environment is so good, the opportunities we get to do the work that I want to do personally is just incredible".
>
> (Interview, 27 June 2014)

Svalbard, and perhaps especially Longyearbyen, is seen as a place of opportunities. A great deal is on offer to develop scientific careers; to undertake research or education in exciting and interesting fields, in a supportive, well-funded and well-equipped way with direct access to the physical phenomena; to influential people in the field; to make connections; and to find others interested in similar work.

Beyond the increasing circulation of value in the sense of political collaboration and funding arrangements, the *practice* of research and its growth creates more value. As networks, collaborations, equipment and experiences expand and link together, the value of doing research in Svalbard in particular spiral outwards in a growing, self-reinforcing cycle of value. Moreover, the glaciers, upper-atmospheric particles and polar bears of Svalbard alike become embroiled in the outwards push for growing knowledge and the value of such knowledge through the specific skills, research methods, equipment and knowledge networks that research in Svalbard generates. As Livingstone notes, place matters in the doing of knowledge production (Livingstone, 2003).

In addition, the enthusiasm, curiosity and other emotional, affectual elements of a relationship with Svalbard as a field site should be considered. While some scientists were reluctant to talk specifically about any form of place attachment or emotional relation to their research, which chimes with previous observations of scientific practices (see for example Lorimer, 2008; Whitney, 2013), equally others at the very least recognised their own enthusiasm and passion for this work and the opportunities that Svalbard presents for them. While the emotional and affectual experiences of scientists and students are usually written out of the formal channels of knowledge circulation and production, they are nevertheless important factors in the practice of research and thus contribute to this growing "value vortex".

"I think all the researchers have an emotional connection to their field . . . having the luck to see glaciers out of my window everyday, . . . it inspires you a lot more. . . . It's amazing we have these images every 10 or 11 days: oh this one is surging; let's go and have a look at it. It's easier to do field work to get the data. But at the same time, we have this very tight connection with the landscape where we are, so I think it's a really big deal for sure".

(Interview, 13 June 2014)

Like others who have observed scientists in the field (Lorimer, 2008; Whitney, 2013), in the growing cyclical relations of value in knowledge production, I noted a common ethics of environmental responsibility, care, respect and affinity for the *wilderness* of Svalbard. This was encouraged in the teachings and practices at UNIS. Such values would seem to be compatible with the environmental goals of the state. However, furthering knowledge and data collection was also given utmost importance among the scientific community. There were clear feelings from the majority of participants that carefully managed access for *all* (including local residents and tourists) should be possible and environmentally beneficial. Reasons for supporting relatively open access included it being an acceptable level of regulation (and therefore keeps good will towards other regulations from the governor); the educational and potential advocacy effects of seeing and experiencing such areas; and the point of view that access should be available equally among citizens. This issue of access is where tensions between different stakeholders surrounding the specific case of the East Svalbard Management Plan coalesced.

Practices of value and values: The East Svalbard Management Plan

"I think that it's much better to do volunteer guidelines, teach people about it so that they can experience it. You protect what you know, and you protect what you can see. You protect what you love".

(Interview, 21 May 2014)

The East Svalbard Management Plan came into force in spring 2014. Though there are no new reserves or parks created as part of the process, the plan tightened restrictions to the access of East Svalbard, turning two zones into scientific reference areas. Other changes were the summer closing of the bird reserve areas to the west of Lågøya and at Tusenøyane, meaning that due to sea ice and restrictions on helicopter usage, it is highly unlikely access will be possible.

Somewhat ironically, the scientific reference area means that access will be limited, even for science activities, the idea being to create an area as unaffected by humans as possible. According to the management plan, climate research and other environmental research that requires access to large and essentially undisturbed areas *may* receive permits for activities. Ongoing monitoring can continue, with the best available environmentally sound technology, but is expected to require less travel and presence in the nature reserves and less direct handling of animals. New surveys that require permits will be kept to a minimum. Surveys that will create basic knowledge about prioritised or red-listed habitats and species, or natural qualities mentioned in the purpose of protection, will be granted permission (Governor of Svalbard, 2012, p. 34).

The new management plan was already taking an effect, with permits being harder to obtain and meet conditions for, as one researcher describes:

> "We've been absolutely slammed recently, by Sysselmannen [the Governor's Office]. . . . they are really trying to limit the areas in which people work. We're very lucky because as geologists we have to go where the rocks are. But I think it's unfair to limit the research in other kind of areas because there's a lot to see over there [East Svalbard]".
>
> (Interview, 27 June 2014)

The management plan was devised through a consultation exercise, lasting over eight years with several different working groups including stakeholders from research, education, and tourism sectors. The starting document from the environment department sought to close off many more areas, but through this consultation process, the stakeholder groups reduced the areas affected by the plan. However, a number of informants observing or participating, felt that the evidence, advice and experience fed into the consultation were not taken on board. They noted that the drawn-out nature of the process itself put stakeholder groups at a disadvantage, given the substantial amount of unwaged time involved. The outcome was still not especially popular and is considered to reflect the values of the Governor's Office and staff and the Norwegian Environment Ministry, often described as "symbolic politics" in action.

> "There has been a case built for protection for this and that, and there was no evidence. All the evidence was flawed, or out-dated, or simply not there, or made on presumption. . . . I must say, I find it hard to take these people seriously now, because they wilfully ignore evidence, which is my profession is a cardinal sin".
>
> (Interview, 3 June 2014)

> "They close one area which is actually quite big which they call a bird reserve: A Thousand Islands [Tusenøyane], yet there's no birds there. . . . you've put lots of lines on the map you made lots of legal texts and operation manuals that no one can understand, but you protect nothing of the environment".
>
> (Interview, 14 May 2014)

Moreover, the physical environment and nonhuman species in this area present a challenge and tension to this approach, whether from not being there, like the birds, or offering unique features, ecologies and conditions that would otherwise be of great interest to tourists and scientists.

Environmental protection has been given the top priority in these eastern zones of Svalbard, despite the predicted increased scientific observations in East Svalbard, which the Svalbard White Paper included (Norwegian Ministry of Justice and the Police, 2010, p. 77). My interpretation is that this is not so much about *what* is valued, because all are in agreement that Svalbard's natural wilderness is highly valuable, but *how* it is valued and what such a valuation does. The same area can have multiple meanings and different kinds of value associated with it: as Endres (2012) puts it, value can be polysemous. In this case, the polysemous nature of value is related to definitions and ideas of wilderness and human relations in and as part of that wilderness.

> "We must have the guts to say no for a lot of activities. Everybody wants to go into the wilderness and they want to go to places which nobody else has been before and we must say no. We don't want people to go in there because of the possibility to destroy the wilderness and the untouched landscape. When you come back in 50 years, it still has to be looking like there hasn't been any people there".
>
> (Interview, 19 June 2014)

We can treat these mountains and glaciers as pristine and an opportunity to show the world high-level environmental protection in action, a time-capsule gift to future generations. The now-familiar critique of the very idea of pristine nature (Cronon, 1995) is, however, recognised by those against the government's approach. Svalbard's unsettled areas constructed as pristine wilderness erase not only the history of natural resource exploitation on and around the archipelago (Avango et al., 2014) but also obscure the present activities. Tourism, fishing, resource exploration, scientific work, environmental protection and geopolitical narratives continue to shape and affect these areas. Those against such an approach might argue that we would never witness the harshness or vulnerabilities of "nature", of our gosling being carried off by the fox, if access were not permitted. Such a fox–gosling encounter might as well not happen as we were not there to learn from it.

> "It looks like in their mindset they [the Governor's Office] are against humans. Humans disturb nature. And nature should be left alone and not disturbed, not visited, and not used by humans. On a philosophical level I disagree. . . . humans are a part of nature, not a separate entity. To uphold this separation is meaningless and indeed harmful".
>
> (Interview, 3 June 2014, and email, 11 September 2014)

On the other hand, valuing Svalbard's wilderness as a site for learning about arctic environments and inspiring wider environmental awareness points to a more open policy of access. Yet this more humanistic approach increases the potential for direct environmental impact and lacks the political impact of declaring an area "fully protected" or closed. Opponents could suggest we would never know how the fox–gosling encounter would play out without human presence, if that presence is always a possibility.

At the Environment Department at the Governor's office, I heard similar views to those complaining about the restrictions: they voiced a will to let people experience "the nature" and therefore inspire protection, *but* they thought that this should occur only in the Western Spitsbergen area, where human settlements and activities have historically been concentrated. It seems that East Svalbard is another matter and is valued in a different way, with Western Spitsbergen acting as a form of a sacrifice zone for tourism, leisure, resource extraction and scientific activities. Taking this view, it matters far less that we potentially disturbed the gosling in Western Spitsbergen than whether we had been elsewhere in Svalbard. Put this way, complex consultation processes, where there are multiple meanings of value at work, can sometimes lead to confusing results:

> "I would say the east Svalbard protection plan probably does damage to the environment. What did you stop there? You stopped 173 environmental protectionists . . . who are some of the richest people in the world, or a film maker or a scientist who wants to do good".
>
> (Interview, 14 May 2014)

Conclusion

Svalbard, as the land of opportunity for scientists, creates economic, social, cultural, environmental and symbolic value in a growing cyclic manner. Scientific engagement with Svalbard also creates *values* that are not necessarily in tune with the values that lie behind the government's approach to environmental regulation. This discord brings the processes of knowledge production into question. Each "camp" makes overt use of quantitative evidence to argue their case, and each accuses the other of being driven by feelings or unscientific practice. However, as Latour (2004) argues, facts and values are not so easily separated. Put another way, Miller (2008) posits that there is a wide spectrum of ways that we value, from quantifiable value (e.g. monetary worth or data sets) to personal values (e.g. care for the Svalbard reindeer or the right to roam freely). Public policies and decision-making processes will achieve better outcomes, Miller demonstrates, if the breadth of different meanings and types of value can be held together and discussed. Indeed, Endres (2012) points to a growing movement in participatory decision-making scholarship that seeks to find a place at the table for polysemous values to be taken into account.

If the values and politics accompanying the scientific evidence drawn on could also be included, perhaps consultation exercises on environmental management in Svalbard could become more satisfactory for all parties. Tracing the value and values associated with knowledge production and its practice in Svalbard has proven to be an example of an illuminating angle for investigation and analysis. I believe that such an approach can open up discussions and make more transparent the processes of evaluation and decision-making at work in contested areas.

Notes

1 I focus largely on research and education activities here; however, tourism shares many characteristics with such activities, including knowledge production (see Saville, 2019).
2 This diversification was also a move to transition Longyearbyen from a mining company town to a more "normal", Norwegian family setting (Grydehøj et al., 2012).

References

Avango, D. et al. (2011) 'Between Markets and Geo-politics: Natural Resource Exploitation on Spitsbergen from 1600 to the Present Day', *Polar Record*, 47(01), pp. 29–39.

Avango, D., Hacquebord, L. and Wråkberg, U. (2014) 'Industrial Extraction of Arctic Natural Resources Since the Sixteenth Century: Technoscience and Geo-Economics in the History of Northern Whaling and Mining', *Journal of Historical Geography*, 44, pp. 15–30.

Cresswell, T. (2012) 'Value, Gleaning and the Archive at Maxwell Street, Chicago', *Transactions of the Institute of British Geographers*, 37(1), pp. 164–176.

Cronon, W. (1995) 'The trouble with wilderness, or getting back to the wrong nature', in *Uncommon Ground: Toward Reinventing Nature*. New York: W.W. Norton & Co, pp. 69–90. Available at: www.williamcronon.net/writing/Trouble_with_Wilderness_Main.html.

Davis, J. S. (2007) 'Scales of Eden: Conservation and Pristine Devastation on Bikini Atoll', *Environment and Planning D: Society and Space*, 25(2), pp. 213–235.

Endres, D. (2012) 'Sacred Land or National Sacrifice Zone: The Role of Values in the Yucca Mountain Participation Process', *Environmental Communication: A Journal of Nature and Culture*, 6(3), pp. 328–345.

Governor of Svalbard (2010) *Svalbard: Experience Svalbard on Nature's own terms* (5th ed.). Longyearbyen: Governor of Svalbard.

Governor of Svalbard (2012) *Management Plan for Nordaust-Svalbard and Søraust-Svalbard Nature Reserves: Draft Proposals for Amendments to the Protection Regulations (English Version)*. Available at: http://oldweb.sysselmannen.no/hoved. aspx?m=44365&amid=3182719&fm_site=44265 (Accessed 17 August 2014).

Graeber, D. (2001) *Toward an Anthropological Theory of Value: The False Coin of Our Own Dreams*. Basingstoke: Palgrave.

Grydehøj, A. (2013) 'Informal Diplomacy in Norway's Svalbard Policy: The Intersection of Local Community Development and Arctic International Relations', *Global Change, Peace & Security*, 26(1), pp. 41–54.

Grydehøj, A., Grydehøj, A. and Ackren, M. (2012) 'The Globalization of the Arctic: Negotiating Sovereignty and Building Communities in Svalbard, Norway', *Island Studies Journal*, 7(1), pp. 99–118.

Henderson, G. (2013) *Value in Marx: The Persistence of Value in a More-Than-Capitalist World*. Minneapolis, MN: University of Minnesota Press.

Henderson, G. L. (2004) 'Value: The Many-Headed Hydra', *Antipode*, 36(3), pp. 445–460.

Jöns, H. (2015) 'Talent Mobility and the Shifting Geographies of Latourian Knowledge Hubs', *Population, Space and Place*, 21(4), pp. 372–389.

Katz, C. and Kirby, A. (1991) 'In the Nature of Things: The Environment and Everyday Life', *Transactions of the Institute of British Geographers*, 16(3), pp. 259–271.

Kings Bay AS (2013) *Kings Bay AS Annual Report 2013*. Ny Alesund: Kings Bay AS. Available at: http://kingsbay.no/kings_bay_as/annual_reports_2/ (Accessed 14 January 2015).

Latour, B. (2004) *Politics of Nature: How to Bring the Sciences Into Democracy*. Translated by C. Porter. London: Harvard University Press.

Lee, R. (2006) 'The Ordinary Economy: Tangled Up in Values and Geography', *Transactions of the Institute of British Geographers*, 31(4), pp. 413–432.

Livingstone, D. (2003) *Putting Science in Its Place: Geographies of Scientific Knowledge*. Chicago: University of Chicago Press.

Lorimer, J. (2008) 'Counting Corncrakes: The Affective Science of the UK Corncrake Census', *Social Studies of Science*, 38(3), pp. 377–405.

Miller, D. (2008) 'The Uses of Value', *Geoforum*. (Rethinking Economy Agro-food Activism in California and the Politics of the Possible Culture, Nature and Landscape in the Australian Region), 39(3), pp. 1122–1132.

Ministry of Foreign Affairs. (2007) *The Norwegian Government's Strategy for the High North, 032191–220013*. Available at: www.regjeringen.no/en/dokumenter/strategy-for-the-high-north/id448697/ (Accessed 29 December 2014).

Ministry of Foreign Affairs. (2014) *Norway's Arctic Policy for 2014 and Beyond – A Summary*, *Government.no*. Available at: www.regjeringen.no/en/dokumenter/report_sum mary/id2076191/ (Accessed 7 January 2015).

Norwegian Ministry of Justice and the Police. (2010) *Report No.22 to the Storting: Svalbard*. Report to the Storting No.22. Norwegian Ministry of Justice and the Police. Available at: www.regjeringen.no/en/dep/jd/documents-and-publications/reports-to-the-storting-white-papers/reports-to-the-storting/2008–2009/Report-No-22-2008–2009-to-the-Storting.html?id=599814.

Norwegian Ministry of Justice and the Police. (2016) *Report to the Storting 32 (2015–2016)* (White Paper No. 32). Retrieved from https://www.regjeringen.no/no/dokumenter/meld.-st.-32-20152016/id2499962/?ch=1&q=.

Pederson, T. (2009) 'Norway's Rule on Svalbard: Tightening the Grip on the Arctic Islands', *Polar Record*, 45(233), pp. 147–152.

Robertson, M.M. and Wainwright, J.D. (2013) 'The Value of Nature to the State', *Annals of the Association of American Geographers*, 103(4), pp. 890–905.

Saville, S.M. (2019) 'Tourists and Researcher Identities: Critical Considerations of Collisions, Collaborations and Confluences in Svalbard', *Journal of Sustainable Tourism*, 27(4), pp. 573–589.

Skeggs, B. (2014) 'Values Beyond Value? Is Anything Beyond the Logic of Capital?', *The British Journal of Sociology*, 65(1), pp. 1–20.

Timothy, D. J. (2010) 'Contested Place and the Legitimization of Sovereignty through Tourism in Polar Regions', in Hall, M. C. and Saarinen, J. (eds) *Tourism and Change in Polar Regions: Climate, Environments and Experiences*. Oxon: Routledge, pp. 288–300.

UNIS (2013) *UNIS Annual Report 2013*. Annual Report. Longyearbyen: University Centre in Svalbard. Available at: www.unis.no/30_ABOUT_UNIS/4010_Root/annual_reports. htm (Accessed 14 January 2015).

Whitney, K. (2013) 'Tangled Up in Knots: An Emotional Ecology of Field Science', *Emotion, Space and Society*, 6, pp. 100–107.

14 Locating value in food value chains

Susan Machum

Industrial food systems are very complex, so recognizing the locations and dimensions of when, where, and how value is created and captured within food supply chains can be challenging. On one hand, "value" has multiple meanings and, on the other, food supply chains are much more multidimensional than standard economic models suggest. This chapter explores the dynamics of food supply chains and the creation of value within them. It argues that at each stage along the food "value" chain are actor groups with competing interests and needs. Thus, researchers undertaking value chain analysis need to consider who is creating value, what the nature of that value is, and who is capturing the value created. Answering these core questions first requires an understanding of the food supply chain, which is the agenda of section one of this chapter. The second section explores the competing value systems implicitly and explicitly embedded within food value chains. It argues that value frames are influenced by both disciplinary interests and by the concerns of particular actor groups and individuals within them. The chapter concludes with reflections on the values that local farmers using organic farming practices use to position and frame their products in the marketplace. This section uses case study data to illustrate that farmers engaged in organic production are often pulled between their commitment to ecological values and their commitment to local production systems.

Dynamics of food supply chains

In the industrialized world, most citizens are purchasing, rather than growing, the food they and their families eat. The capacity to walk into and select food from a grocery store lined with shelves of items from places far and wide is evidence of a well-organized and highly efficient global food system. However, this capacity to source food from farmers thousands of miles away was not always the norm. Historically, people grew their own food and exchanged excess with neighbouring farmers, and those who didn't grow food relied on local farmers to supply their tables. Over time, the distance between food consumers and food growers has expanded exponentially; this "distancing" (Kneen, 1995) has resulted in a global food system where in rural Canada it is possible to find apples grown in South

Africa sitting alongside locally grown varieties. What's more, the imported apples are often cheaper than those grown locally.

Tracking the movement of goods and services from their point of creation to their point of final sale is referred to as supply chain research. Typically, economic supply chains document the production, distribution, and consumption of goods and services. In the case of food systems, it has been standard practice to add a step and investigate how foods are grown, *processed*, distributed, and consumed. This separation of food growing from food processing – when in fact both are dynamics of food production – reflects the industrialization of food and the significant shift away from eating "raw" foods in the form of meat, eggs, vegetables, and other "simply" processed foods, such as dairy products (milk and yoghurt) and bread, to the growth in highly refined, processed foods that Winson (2014) calls pseudofoods. The differentiation between production and processing also highlights the considerable weight that processing industries have come to wield in the world of food. More recently, food chain research has added *waste* as a dynamic of the production-consumption cycle, but waste is usually added on at the end of the whole cycle. Thus, food supply chains generally refer to production, processing, distribution, consumption, and *waste* cycles.

This framing of supply chains as neat, orderly, successive stages belies how complicated and messy food supply chains really are. Take waste, for example. Waste doesn't just appear at the end of the process; it is seeping out along the entire supply chain. In fact, a major agenda of value chain management has been to reduce and repurpose waste all along the supply chain in order to improve the bottom line (McDonough and Braungart, 2002). But rather than address how waste flows throughout the entire cycle, the primary focus of most food waste research has been on individual and institutional "consumer" food waste, resulting in research on waste in the family household, restaurants, and other institutional kitchens at the expense of other sites along the supply chain. Perhaps this approach of adding waste to the end of the cycle and effectively turning it into a post-production consumer problem emerges from the notion that what consumers do with the excess packaging, leftovers, and other food waste generated has little direct impact on the corporate bottom line. The flaw in this logic emerges from the realization that waste is present at every stage. Consequently, McDonough and Braungart (2002) argue that for the health and well-being of the planet's ecosystem, corporations, rather than individuals, should be responsible for a product's development and the waste it generates from its point of conception to its point of obsolescence.

In practice, rather than being separate and distinct phases, waste, consumption, production, processing, and distribution are features of each stage of the food supply chain. As Figure 14.1 illustrates, farmers are typically identified as primary *producers*. However, they are also *consumers* of seed, fertilizers, fuel, tractors, ploughs, combines, and other technological inputs; *distributors* of the goods they produce; creators of waste; and more and more often engaged in on-site *food processing*. As consumers, the resources that farmers purchase are the necessary

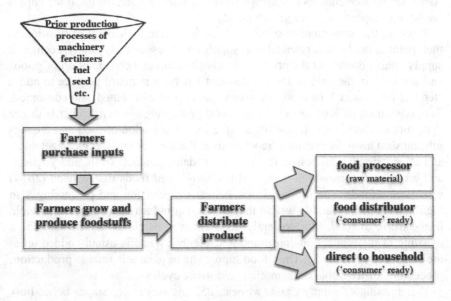

Figure 14.1 Farmers' complex roles in the food value chain

ingredients or raw materials for growing and harvesting their foodstuffs. Likewise, food-processing companies need a significant number of inputs – including raw foodstuffs – to undertake their production activities, and they too must act as consumers of these inputs to accomplish their tasks. Thus, both farmers and processing companies are simultaneously consumers and producers, even though they are often framed solely as producers in the food supply chain, while the purview of consumption is usually relegated to the humans (rather than the processing machines) who will physically *ingest* the food item.

It is not just food that is being moved within the food supply chain but also the equipment, energy, labour, and resources needed to transform and transport "raw materials" into finished goods. In order for each sector to buy the industrial inputs that they need for their production processes, these goods must be distributed into the marketplace by other producers. The length of the distribution network will vary from input to input; some products may travel far, while others may be sourced locally. Moving goods around in the marketplace is an integral part of the whole supply chain, not a single intermediary step, as it is often portrayed in supply chain models.

Even distributors are consuming goods and services to carry out their role in the supply chain: they purchase vehicles, gasoline and diesel, trolleys, pallets, and so on to handle and move items from point A to point B. What distributors "produce" is a service: they bring the raw materials and finished goods into the physical domain of the company or person who needs or wants them. By relieving individual and industrial "consumers" of the need to fly or drive to the source of potato production or farmers to the location of tractor manufacturing, distributors expand the sales domain of each producer and help match potential consumers to the "finished" products they want. The steps involved in distribution can be quite complex, as Albisu et al. note:

> Even in fresh agricultural product chains, which involve minimal processing, a large number of activities such as refrigeration, packaging, logistics, etc. are undertaken before the products of farmers and agri-food enterprises reach the market, either through the intermediary of wholesalers or retailers or, in certain circumstances, directly to the final consumer.
>
> (2010: p. 252)

Gereffi et al. (2008, p. 359) claim the input–output structures of value chains should encapsulate "the entire process that brings a product or service from initial conception to the consumer's hand. The main components typically entail research and development (R&D), production, distribution, marketing, and sales". However, their framing of each step as linked to a previous node nevertheless leaves each node theoretically and practically distinct. This obscures the reality that producers are also consumers and that consumers are also producers. This latter point is well made by Delphy and Leonard (1992), who contend that we must be eating "raw chicken out of greengrocers' bags" if we are not processing and effectively "adding value" by cooking the partially prepared products that we bring into the family household. In truth, most food purchases require some cooking, assembly, or reheating before we can ingest them – that is, not all food is highly processed and ready to eat straight out of the package.

Despite the complexity in each node, tracing the movement of food from production to consumption (i.e. the raw material stage to its final point of ingestion) has been the key agenda of food supply research. In recent decades, this has often meant studying corporate takeovers and the horizontal and vertical integration of corporate food giants. Gereffi et al. (2008) provide two excellent case studies of how corporations gain control of chicken and tomato supply chains, and an historical study by Senopi illustrates how McCain Foods International cornered all dimensions of the French fry market in their headquarters' province of New Brunswick, Canada:

> Potatoes are grown on McCain land (Valley Farms Ltd) enriched by McCain fertilizer (McCain Fertilizers Ltd) using McCain seed (Foreston Seed Co. Ltd). Harvesting is done with McCain machinery (Thomas Equipment Ltd) and the harvested potatoes are either stored in McCain facilities (Carleton Cold Storage Co. Ltd), sent to McCain's plant for processing (McCain Foods Ltd) or sold fresh. In the latter case, the potatoes are handled by McCain shippers (McCain Produce Co. Ltd) which use McCain trucks (Day and Ross Ltd) to move them to McCain storage facilities (Bayside Potato Port Ltd) at the point of shipping. The processed potatoes can similarly be moved in McCain trucks (M. & D. Transfer Ltd) for shipment abroad where one of

McCain's sales distribution systems (McCain International Ltd) handles the marketing.

(Senopi, 1980, pp. 34–35)

What the three examples illustrate is how one corporation successfully comes to own and link various companies in all the "backward and forward linkages, up to the level in which the raw material is linked to the final consumers" (Nang'ole et al., 2011, p. 1). This process is referenced in agri-food supply chain research as vertical coordination (Hobbs and Young, 2000) or "as a vertical alliance or strategic network between a number of independent business organizations within a supply chain" (Nang'ole et al., 2011, p. 1). Ironically, this vertical axis is almost always presented graphically along a horizontal axis (see Hobbs and Young, 2000, p. 133). The horizontal axis captures the relationships among actor groups on the same rung of the ladder (Nang'ole et al., 2011, p. 1), whereas the vertical axis marks the relationships between various stages of the food supply chain. Specifically, horizontal relationships are those between farmers in producer cooperatives or farmers' markets or between consumers in buying cooperatives or corporate takeovers in the same industry.

Increasingly, vertical integration is allowing global food corporations to control the entire supply chain – including, in many cases, the everyday work practices of farm operators. For example, McCain Foods International – and many other food-processing corporations – pursue formal production contracts with farmers that establish how the farmers will grow and store their crops. These agreements are increasingly turning farmers into little more than corporate "employees" (Hobbs and Young, 2000; Gereffi et al., 2008, Senopi, 1980). Farmers who sign pre-production contracts have little autonomy over when, how, and where to market their product, whereas those without pre-production contracts have a lot more independence and autonomy. It is only farmers without pre-production contracts who can, post-production, assess the market and opt to sell their products to food processors and food wholesalers or directly to household consumers.

Unpacking how the global food system is evolving and the roles and relationships embedded in it is therefore challenging. First, there are a series of actors within any food system – producers, processors, distributors, consumers, financiers, policymakers, and so on – who all engaged in multiple roles. Second, each one of these groups in the supply chain has its own agenda and aims to promote its own interests. Third, in each actor group, individuals embody different value systems, each of which promotes distinct forms of action and the evaluation of ongoing practices. It is interest in these competing agendas and values that has prompted a shift in the discourse from food *supply* chains to food *value* chains.

Framing value

Rather than focus on the step-by-step actions required to move product from point of conception to point of consumption, food value chains evaluate the kinds of social relationships embedded in each transaction point. For example, Allen

(2006) charts the distinctions between the two discourses in terms of their language and focus of concern; she indicates that supply chains study commodities and prices, while value chains focus on product differentiation and the type of value captured. On the other hand, Nang'ole et al. (2011) argue that the two phrases remain interchangeable despite the important differences. In practice, the shift from a *supply chain* to a *value chain* discourse began with interest in how, when, and where *economic value* was being added in the production and distribution of food. But the growing local food movement is expanding the debate by questioning the *kinds of value* embedded in the food we eat. They want to know whether our food choices harm or sustain local ecosystems, support local food farmers or large corporate interests, or ameliorate or exacerbate levels of social inequality.

Given the propensity of social science disciplines to promote different ideological frameworks, how researchers and practitioners perceive and evaluate 'value' within the food system will vary according to disciplinary training. For example, economists are generally concerned with the dollar values added by different transactions along the supply chain. Economists calculate returns on investment and refer to exchange value – that is, the amount of money earned when goods and services are sold within the marketplace. An economist investigating the food system will ascertain where financial losses are taking place and make recommendations to reduce such losses through new production processes and innovation. Their focus on revenue streams and profits supports product development, corporate amalgamation, and integration schemes. From the economist's vantage point, value is a cost-price issue wherein each seller will be aiming to sell high and every buyer will be trying to find the best bargain (Redmond, 2013). In this case, the best way for corporations to maximize profits is to follow McCain's lead and pursue vertical and horizontal integration (Senopi, 1980; Ibrahim, 2007; Howard, 2016).

Meanwhile food nutrition scientists focus on the nutritional value of particular foods, and they recommend or discourage the consumption of foods based on how they reflect recommended daily allowances. Nutritionists argue that nutritionally dense foods high in fibre have more food value than those filled with empty calories. Often, they are advocates of food labels since labels help consumers make informed decisions about the health benefits or drawbacks of eating particular foods. In practice, the nutritionist's value system usually stands in sharp contrast to the economist's value system because "whole" or raw foods generally offer high nutritional value but low economic returns compared to "pseudofoods" that have low nutritional value but relatively higher economic returns.

Conversely, cultural anthropologists concentrate on the value of food traditions, food ways, and rituals for reproducing cultural practices (Anderson, 2014). In fact, cookbook sections of bookstores are overflowing with advice and information on how to replicate and capture cultural traditions. What we eat, when and where we eat, and how we prepare foods are all indicative of the historical moment and cultural context we find ourselves in. Preparing, cooking, and eating food in ways indigenous to our cultural traditions is valuable for retaining or transforming culture. As Roe (2006, p. 117) elaborates, "The symbols, messages and meanings . . . of [our] practical engagement [with food] arises from material relations, consisting of what is 'afforded' between the environment and the human". It is in this context that Crowther (2013) argues that we "eat culture" through the food we opt to ingest.

Ethicists are concerned with the value systems that are in place for retaining honesty, truth, and justice in social relationships. Their examination of the food system leads them to consider how ethical work relationships are along the supply chain, how fair the division of labour and riches are, and the type and quality of interactions among actor groups within the supply chain (see Roe et al., 2005). Ethicists and environmentalists both examine how sentient beings treat non-sentient beings – specifically they examine human–animal relations and human–plant relations. Embedded in this ethics environment, society-nature debate is the environmentalist's concern with the value of nature, the value of diversity, and ultimately the value of retaining a "healthy" ecosystem.

Like ethicists, sociologists are concerned with understanding competing value systems. We are particularly interested in how different actor groups within the food system construct and interpret value; we investigate who benefits and who loses from existing social, economic, and cultural practices, and we recognize that actor groups often have competing agendas. We ask who is winning and losing from the existing order – in this case, who has the resources to create, distribute, and retain the "value" created along the food value chain. Sociological research thus tends to focus on an individual's specific *location* within the food *system*. We recognize that individual food consumers often have a different location and agenda from farmers and food processors. In this vein, a significant amount of food value chain research focuses on the experiences of either food consumers or food producers – which is sometimes referred to as buyer-driven versus producer-driven models (Gibbon et al., 2008).

According to Dagevos and Van Ophem (2013), a consumer-centred approach to value captures four distinct value frames: product value, process value, location value, and emotional value. Product value refers to both the nutritional and economic value of the food item, whereas process value takes into account the ethical practices and processes of food production. Referencing the work of Lusk, they explain that food products have 11 general attributes that can be explored in depth to appreciate the dynamics of food value: "safety, nutrition, taste, price, convenience, appearance, naturalness, tradition, environmental impact, fairness and origin" (Dagevos and Van Ophem, 2013, p. 1481). The first six relate directly to product value, and the last five are more relevant to process value. Process value could also encompass questions about the gendered division of labour within the food supply chain (Barrientos, 2001), the ecological sustainability of production practices (Renting et al., 2003), and the kinds of relationships between producers and consumers that particular food systems promote (Starr, 2010; Marsden et al., 2000). Directly related to this last point, location value acknowledges the context and setting in which the food is purchased or consumed. Location value is thus, for many researchers, a critical dimension of process value. Finally, emotional value involves the consumers' feelings about the food, the brand, and the overall "food" experience. In practice, product value, process value, location value, and emotional value are more widely discussed in the health literature in terms of the acceptability, availability, affordability, and accessibility of food for consumers within the marketplace (Gereffi et al., 2008; Caspi et al., 2012). Furthermore, all of these value schemes are used to assess both industrial *and* alternative food supply chains.

Both consumer-centred value frames recognize that whether they are in their local farmers' market or mega-chain grocery store, when consumers are selecting items to put in their shopping basket, they are consciously and unconsciously deciding whether or not the foods before them meet particular standards, including environmental, moral, and ethical ones. Of course, what goes into the cart will depend on what is available in the community, the financial resources the consumer has, their dietary needs, their personal tastes, and their cultural practices. So while the food itself may contain particular nutritional and sociocultural values, the value of that food to consumers will vary depending on the individual.

On the other end of the food supply chain, farmers and processors are seeking personal validation and financial reward for their efforts. Their goal is to capture as much of the consumer dollar as possible, so to sell at higher prices, they are engaged in "adding value" to food products as they move upstream. Producer-focused research thus examines how the food dollar is distributed in the marketplace or more specifically who is able to "capture" the created value of a particular good or service (Skilton, 2014). What this food value chain research is investigating is the complex relationships between various "producer" groups in the design, production, marketing, and distribution of foodstuffs (Allen, 2006; Gereffi et al., 2008; Gibbon et al., 2008; Young and Hobbs, 2002).

While most of the research on food value chains has emphasized either corporate efforts to improve and manage resources more effectively or consumers' food experiences, "value" is increasingly being reframed to acknowledge ecological, geopolitical, and social justice issues. It is recognized that every consumable was produced by human labour using natural resources. From this framework, the values in every consumer good are extensive: Were production practices ecologically sustainable (Renting et al., 2003; Marsden et al., 2000)? Are the foods free of pesticides, other additives, and GMOs (Winson, 2014)? Were working conditions favourable, and were employees fairly compensated for their contribution (Barrientos, 2001; Gereffi et al., 2008; Brown, 2013)? Is food preparation inside the family home, and in particular, is women's reproductive labour recognized as an essential contribution to the food supply chain (Delphy and Leonard, 1992)? As we answer these questions, we come to see the food we eat as a dynamic entity filled with multiple layers of decisions and actions, all of which embody particular value systems.

Building alternative food value chains

It is this recognition that consumables are filled with layers of prior decisions that either implicitly or explicitly support or undermine core values that has spurred ecologically and socially "conscious consumerism" (Brown, 2013; Jones, 2008). Alternative food networks (AFNs) are emerging as part of this larger movement in an effort to better trace and document the pathways our food travels from farm to plate (Barrientos, 2001; Brown, 2013; Gottlieb and Joshi, 2013; Marsden et al., 2000; Renting et al., 2003; Starr, 2010). Phrases typically used to capture this process are "farm to table", "farm to fork", and "gate to plate"; my preference is "seed to soil" since it evokes a more comprehensive cycle of action, whereby we recognize that all organic matter is eventually returned to the soil.

Of course, to fully comprehend and document every action, its ideological framework, and underlying value system embedded in the products we buy would be mind-bogglingly complex. Nevertheless, food traceability programmes aim to do just that "because the information about the product – its origin, the management practices, or other attributes – must stay with the product from production to end consumer for this value to be fully realized" (Allen, 2006, p. 24). Without this documentation, value claims cannot be verified. The purpose of third-party fair trade and organic certification programmes is to provide the consumer with independent verification that their food choices meet ecological and labour standards (Brown, 2013). They are a means for the consumer to determine the authenticity of the "value" claims embodied in the food they intend to consume. Most often, consumers rely on product labelling to determine whether the food is organic, local, GMO-free, marketed under fair trade conditions, and so on. But the agenda of AFNs is to go a step further and through the creation of short food supply chains (SFSC), rebuild local food systems and relationships that have been systematically eroded through the development of longer and longer food supply chains.

Just how weak our relationship is with local food producers became apparent in New Brunswick a few years ago when ACORN (the Atlantic Canadian Organic Regional Network) ran an effective campaign asking locals what the name of their farmer was. It began by asking, What's the name of your doctor? Who's your dentist? And it went on to ask, What's the name of your hairdresser? Who shovels your driveway in the winter? Who mows your lawn in the summer? As images of people engaged in these tasks flew up, most had ready answers. But when asked, Who grows your food? our minds drew a blank. Even though food is an intimate part of our daily lives, global food markets have stripped away and systematically eroded the relationships between farmers and food consumers.

In practice, SFSCs challenge the marketing and distribution strategies of industrial agriculture, which hide and disguise the details of food production, processing, and distribution. In short-supply chains, rather than purchase food at impersonal box stores, consumers meet those who grow their food at farmers' markets, at the farm gate, and through community-supported agriculture (CSA) initiatives. By promoting personal contact, these marketing techniques help create a rapport between growers and buyers (Starr, 2010; Marsden et al., 2000; Renting et al., 2003). Another advantage is that they allow "the consumer to make value judgments about the relative desirability of foods on the basis of their own knowledge, experience, or perceived imagery (of the food)" (Marsden et al., 2000, p. 425). Through direct contact, the producer can disclose the conditions under which the food was grown: often on the family farm with family labour and/or using organic farming practices. Being able to regularly meet a local farmer builds personal awareness of where food is coming from and the conditions under which it was produced; from all accounts, such direct contact turns into a win-win situation for local food producers and consumers (Marsden et al., 2000; Starr, 2010).

A recent turn in the AFN has seen consumers and producers struggle over what is more important to support – local food production or organic food production. Activists in the AFN are concerned about many issues, including animal welfare, food miles, ecological degradation, how renewable energy sources are, and the rate of energy consumption. While many believe organic food production is the critical way forward, those promoting local cultural and economic stability favour local products. The debate between organic and local emerged in part because it was unclear whether or not supermarket organics were a true alternative to industrial agriculture and its questionable production practices (Pollan, 2008). At the base of SFSCs are local food systems that embody a fundamentally different set of values than industrial food value chains. As Starr explains,

> The "*cosmology*" of the local food movement is food as community (instead of commodity). The movement aims to build "local food systems" based on ecological analyses such as watersheds, sustainable farming, seasonality, heritage of biodiversity, and cultural preferences. Food is transformed from a commodity to a pleasure made possible by human relationships.
>
> (2010, p. 484, original emphasis)

SFSCs implicitly support the agenda of food sovereignty in that they seek to give farmers and consumers control of their local food system. As Wittman et al. (2010, p. 2) explain, "broadly defined [food sovereignty] is the right of nations and peoples to control their own food systems, including their own markets, production modes, food cultures and environments". Controlling the food supply chain allows those within it the opportunity to decide the values that should or should not be supported.

Yet a prevailing question is, what should be prioritized, local or organic production? The two concepts are not synonymous. Local generally embodies economic values, while organic captures ecological ones. Looking at the interplay between local/global and organic/industrial, four options emerge: one can buy food that is produced locally using industrial farming practices; one can buy foods grown locally using organic practices; one can purchase foods grown organically but embedded in a global supply chain; or one can buy industrial foods distributed through a global supply chain. Each option supports different value systems. Those who want to support local economies and ecosystems will argue that local organic is the best option. However, choices imply first that all options are equally available and second that people have the economic means and resources to pursue them. Such is not always the case; for example, extensive research exists on the dynamics of social inequality and their impact on food availability, accessibility, affordability, and acceptability decisions.

As food consumers, every time we go shopping, we are making choices that implicitly or explicitly support particular food value chains. To illustrate, I live 150 km from the original headquarters of McCain Foods International, where potatoes continue to be processed into French fries. This corporation does not buy organic potatoes, but many of the products they produce are locally grown and processed using conventional industrial practices. McCain Foods is clearly a major player in the global food system; but given my geographic location, every time I make a purchase from their company, I am supporting a "local" enterprise. However, when organic trumps local, I can easily find myself buying fruits and vegetables that have been produced on an industrial scale, been picked by migrant workers, and travelled 5,500 km from California. With these choices, it is hard to know which pathway is better. Choosing local organic is much easier during the height of the summer growing season. Given that I live in Eastern Canada, at a personal level, staying committed to local organic requires a radical shift in food preservation and storage practices at home, something that has waxed and waned depending on my other commitments and responsibilities.

As food producers, farmers also struggle in their efforts to pursue particular pathways. Recently, I supervised an undergraduate honours thesis that explored the rationale farm families use to make the decision of whether or not to formally opt in or opt out of organic certification programmes (Dawe, 2015). What we had been observing is that a number of farm families use organic farming practices but avoid formal certification. They are engaged in a significant number of the marketing strategies outlined by the AFN, in that they are selling at local farmers' markets, directly from the farm to consumers, and through CSAs. And it is through this personal contact with their customer base that they are able to share information about their growing techniques and use of organic versus synthetic compounds. Of note, those who have opted to certify are incredibly committed to the third-party verification process and suspicious of the claims to farming organically of those who have not sought such verification. As Dawe notes, respondents who have opted to certify insist that "certification assures consumers that their food has met a minimum level of being chemical free, fertilizer free, and that it has been grown to meet certain standards" (2015, p. 99). Certification is thus a valuable marketing tool that allows farmers to charge premium prices. In addition to verifying the ecological soundness of their product, formal organic certification programmes allow farmers to gain economic benefits. Third-party organic certification is thus a strategy for gaining competitive advantage, especially for those who want to market their products outside of the local area.

In contrast, those who have opted out of formal certification processes emphasize that they are not necessary for building producer–consumer relationships and selling locally. One participant in this category emphasized that through one-to-one contact with customers, he is able to communicate "his products are pesticide- and GMO-free, . . . which attracts consumers demanding organic vegetables" (Dawe, 2015, p. 86), despite the lack of third-party verification. Key factors for not pursuing third-party certification are the breadth and scope of available programmes, inconsistency in standards, and the additional costs to the family farm for undertaking certification, such as initial costs, annual fees, and the ongoing paperwork to retain certification. In effect, the costs of gaining economic advantage down the road come at a financial cost and a time and energy cost in the present (Dawe, 2015). However, not participating in the certification process means farmers "spend a great deal of time explaining their practices to consumers" (Dawe 2015, p. 103). Such explanations are obviously not possible if a product is sold via a third-party. In addition, despite growing interest in regional cuisine and *terroire*, growing and selling "local" products is nontransferable from the physical location in which it is produced, whereas "certified organic" is a label that travels across time and space. In short, Dawe's (2015) case study suggests there is no real substitute for local.

In this example, we are looking at the challenges of capturing ecological and economic value from the perspective of farm producers. When we bring in consumer values of taste and price preferences or add cultural, nutritional, political, and other values, our assessments can start to get unwieldy. Nevertheless, we need to start somewhere. How farmers seek to measure, embed, and convey organic practices and the "localness" of their products to potential consumers illustrates the challenges they face in wedding their own value systems with their customer base. The AFN and local food movements are promoting a dramatically different value set than the economic efficiency and profitability model in much value chain literature. Despite ongoing efforts in the ecosystems service and community sustainability literatures to do so, it remains difficult to quantify and measure the value of retaining an ecosystem or the value of building local community. Yet this is the precise agenda of food activists promoting food value chains (see Allen, 2006).

As researchers, we need to recognize the way that language, discourses, and practices are entangled in double and triple meanings. Agricultural economists and grocery store chains have a different agenda for food value chain research than do local farmers and activists. On a daily basis, values – whether they are economic efficiency, particular cultural practices, biodiversity, or fair wages and good working conditions – are being woven into each and every food system. Food traceability programmes aim to help us understand what those values are in a given food item, but it is our job as consumers to decide whose production practices we will endorse when we fill our shopping baskets. However, as Waxman (1996) notes, our choices are really limited to what is available. To support an alternative value chain, alternatives must be available on the supermarket shelf. Consumer sovereignty must be accompanied by collective action, changes in food policies, and government regulation to seriously constrain and challenge the industrial food system (Carolan, 2017).

Conclusion

This chapter has argued food supply chains are not a simple step-by-step process. Multiple layers, geographies, and scales of production are constantly at play. At every node along the "vertical" food value chain, several forms of activity with vastly different ecological, labour, political, cultural, and social decisions are being pursued. Our efforts to simplify processes and comprehend them as upstream or downstream, vertical or horizontal, or driven by buyers or producers obscure layers of complexity within the food system. This chapter has sought to explore this complexity and highlight the challenges that farmers face as they build ecologically and locally sound alternatives to the industrial food system.

References

Albisu, L., Henchion, M., Leat, P. and Blandford, D. (2010) 'Improving Agri-food Chain Relationships in Europe: The Role of Public Policy', in Fischer C and Hartmann M (eds) *Agri-food Chain Relationships*. Wallingford: CABI, pp. 250–266.

Allen, J. (2006) *Assessing the Market Dynamics of 'Value-Added' Agriculture and Food Businesses in Oregon: Challenges and Opportunities*. Portland State University: Center for Sustainable Processes and Practices.

Anderson, E. (2014) *Everyone Eats: Understanding Food and Culture* (2nd ed.). New York: New York University Press.

Barrientos, S. (2001) 'Gender, Flexibility and Global Value Chains', *IDS Bulletin*, 32(3), pp. 83–93.

Brown, K. (2013) *Buying Into Fair Trade: Culture, Morality, and Consumption*. New York: New York University Press.

Carolan, M. (2017) 'Agro-Digital Governance and Life Itself: Food Politics at the Intersection of Code and Affect', *Sociologia Ruralis*, 57(S1), pp. 816–835.

Caspi, C., Sorensen, G., Subramanian, S. and Kawachi, I. (2012) 'The Local Food Environment and Diet: A Systematic Review', *Health & Place*, 18(5), pp. 1172–1187.

Crowther, G. (2013) *Eating Culture: An Anthropological Guide to Food*. Toronto: University of Toronto Press.

Dagevos, H. and van Ophem, J. (2013) 'Food Consumption Value: Developing a Consumer-Centred Concept of Value in the Field of Food', *British Food Journal* 115(10), pp. 1473–1486.

Dawe, L. (2015) *To Certify or Not to Certify? The Decision of Vegetable Farmers in New Brunswick to Opt In or Opt Out of Organic Certification Programs*. Unpublished BA Honours thesis: Department of Sociology, St. Thomas University, Fredericton, NB.

Delphy, C. and Leonard, D. (1992) *Familiar Exploitation: A New Analysis of Marriage in Contemporary Western Societies. Oxford: Polity Press*.

Gereffi, G., Lee, J. and Christian, M. (2008) 'US-based Food and Agricultural Value Chains and Their Relevance to Healthy Diets', *Journal of Hunger & Environmental Nutrition*, 4(3–4), pp. 357–374.

Gibbon, P., Bair, J. and Ponte, S. (2008) 'Governing Global Value Chains: An Introduction', *Economy and Society*, 37(3), pp. 315–338.

Gottlieb, R. and Joshi, A. (2013) *Food Justice*. Cambridge: The MIT Press.

Hobbs, J. and Young, L. (2000) 'Closer Vertical Co-ordination in Agri-food Supply Chains: A Conceptual Framework and Some Preliminary Evidence', *Supply Chain Management* 5(3), pp. 131–142.

Howard, P. (2016) *Concentration and Power in the Food System: Who Controls What We Eat?* London: Bloomsbury Publishing.

Ibrahim, D. (2007) 'A Return to Descartes: Property, Profit, and the Corporate Ownership of Animals', *Law and Contemporary Problems: Animal Law and Poli*cy, 70(1), pp. 89–115.

Jones, Ellis. (2008) *The Better World Handbook Shopping Guide* (2nd ed.). Gabriola Island, BC: New Society Publishers.

Kneen, B. (1995) *From Land to Mouth: Understanding the Food System* (2nd ed.). Toronto: NC Press Ltd.

Marsden, T., Banks, J. and Bristow, G. (2000) 'Food Supply Chain Approaches: Exploring Their Role in Rural Development', *Sociologia Ruralis*, 40(4), pp. 424–438.

McDonough, W. and Braungart, M. (2002) *Cradle to Cradle: Remaking the Way We Make Things*. New York: North Point Press.

Nang'ole, E., Mithöfer, D. and Franzel, S. (2011) *Review of Guidelines and Manuals for Value Chain Analysis for Agricultural and Forest Products*. ICRAF Occasional Paper No. 17. Nairobi: World Agroforestry Centre.

Pollan, M. (2008) *In Defense of Food*. London: Penguin Books.

Redmond, W. (2013) Three Modes of Competition in the Marketplace. *American Journal of Economics and Sociology* 72(2), pp. 423–446.

Renting, H., Marsden, T. and Banks, J. (2003) 'Understanding Alternative Food Networks: Exploring the Role of Short Supply Chains in Rural Development', *Environment and Planning*, 35(3), pp. 393–411.

Roe, E. (2006) 'Things Becoming Food and the Embodied, Material Practices of an Organic Food Consumer', *European Society for Rural Sociology*, 46(2), pp. 104–121.

Roe, E., Murdoch, J. and Marsden, T. (2005) The retail of welfare-friendly products: A comparative assessment of the nature of the market for welfare-friendly products in six European countries, in Butterworth, A. *Science and Society Improving Animal Welfare, WelfareQuality® Conference Proceedings*, Brussels, NP, pp. 35–40.

Senopi Consultants Ltd (1980). *A Report on the Situation of New Brunswick Potato Farmers for the National Farmers Union*. Saskatoon: NFU Mimeo.

Skilton, P. (2014) 'Value Creation, Value Capture, and Supply Chain Structure: Understanding Resource-Based Advantage in a Project-Based Industry', *Journal of Supply Chain Management* 50(3), pp. 74–93.

Starr, A. (2010) 'Local Food: A Social Movement?', *Cultural Studies < – > Critical Methodologies*, 10(6), pp. 479–490.

Waxman, N. (1996) 'Cooking Dumb, Eating Dumb', in Washburn, K. and Thornton, J. (eds) *Dumbing Down: Essays on the Strip-mining of American Culture*. New York: W.W. Norton, pp. 297–307.

Wittman, H., Desmarais, A. A, and Wiebe, N. (2010) (eds). *Food Sovereignty: Reconnecting Food, Nature and Community*. Halifax: Fernwood Publishing.

Winson, A. (2014) *The Industrial Diet: The Degradation of Food and the Struggle for Healthy Eating*. New York: New York University Press.

Young, L. and Hobbs, J. (2002) 'Vertical Linkages in Agri-food Supply Chains: Changing Roles for Producers, Commodity Groups, and Government Policy', *Review of Agricultural Economics*, 24(2), pp. 428–441.

15 Private finance evaluation among REDD+ projects in Indonesia

Rowan Dixon

Introduction

Forests have recently attracted attention from investors seeking capital gains from a novel commodity that they appear to embody. Not traditionally the concern of international finance, the carbon stored in these forests has emerged as an investable object. Following global efforts to reduce greenhouse gas emissions, the conservation of forest carbon has been touted to deliver capital profits based on carbon markets. Markets that facilitate businesses offsetting their carbon emissions by purchasing units that represent and finance the conservation of forests that would otherwise be lost, along with their stored carbon. The intervention in this case is REDD+ (reducing emissions from deforestation and forest degradation) through the sustainable management of forests and the conservation and enhancement of forest carbon stocks (UN-REDD, 2015).

REDD+ is a real-world exercise in '*putting a value on nature*' (Sukhdev, 2011), aiming to revalue and reconstitute forests as bundles of commodities that can be invested in. These bundles allure profit-seeking finance (PSF) to invest for exchange and accumulation. This chapter draws on the author's PhD research to do two things. It first introduces the REDD+ processes of commodification and market environmentalism in their global contexts of financial exchange. Second, it introduces how the global value chain (GVC) framework, in partnership with social relations of value (SRV) theory, can aid attempts to locate value in REDD+ projects in Indonesia and their exchange.

Despite supposed coherence as a global programme – championed by the United Nations Collaborative Programme on REDD+ (UN-REDD) and the World Bank's Forest Carbon Partnership Facility (FCPF) – REDD+ is emerging in highly contingent ways. As part of wider voluntary carbon market mechanisms, REDD+ has developed in a decentralised fashion, with minimal formal regulation, and is confronting numerous methodological obstacles. These voluntary forest carbon markets evolved in relation to other, much larger compliance markets, most notably the clean development mechanism (CDM). The process of consolidation and regulation for REDD+ brought about increasingly reputable voluntary standards that legitimise REDD+ projects. However, REDD+ offsets remain a small part of the voluntary market, accounting for around 9% of transactions (Peters-Stanley and Yin, 2013).

The future costs of addressing climate change are unclear, but the United Nations Convention on Climate Change predicts that new and additional investment of around 0.3%–0.5% of global GDP per annum is required by 2030 (UNFCCC, 2007). These preliminary figures have since been described as considerably underestimated (Fankhauser, 2010) and likely to outstrip the resources available to public sources, positioning PSF as the last bastion with sufficient finance (Ervine, 2013). This desired co-option of PSF has been accompanied by international favour for "flexible mechanisms" for emissions trading and offsets – increasing attention on optimising markets and incentivising the shift of global private investment patterns towards mitigation and adaptation activities (UNEP-FI, 2014).

Despite the slow progress in developing the mechanism, significant public sector resources, about US$3 billion, have been channelled into REDD+ (Nakhooda et al., 2013). This has focused primarily on research and capacity building to help forested countries attain REDD+ readiness through establishing institutions, strategies and practices in anticipation of compliance markets. Subsequently, a diversity of REDD+ projects has been piloted in a number of developing countries, driven by diverse PSF actors (Dixon and Challies, 2015).

Commodification in market environmentalism

As an example of market environmentalism, REDD+ seeks to govern resources via capitalist market exchange while optimising economic, social and environmental ends (Anderson and Leal, 2001). Thus, aligning with the pursuit of "sustainable development" (Rogers et al., 2008), or "green growth" (OECD, 2011), to re-embed and "renature" the economy (Dunlap, 2015). This establishes a "permissive normative context" of contemporary neo-liberalism (Paterson et al., 2013, p. 3) where climate change is mitigated by transforming ecosystem services and pollution into commodities and derivatives to be traded on financial markets and permitting the structural power of capital interests in policy discourse and practice (Ervine, 2013).

Lövbrand and Stripple (2012, p. 671) observed "indirect regimes of calculation" to commodify carbon as exacerbating financial rationales; where carbon markets "become popular because they have enabled businesses to imagine a cycle of investments, profits and growth" (Paterson, 2012, p. 83), shifting the focus from *whether* and towards *how* PSF might facilitate REDD+ (Hiraldo and Tanner, 2011). This accompanies a reliance on consumers' actions of "purchasing" green commodities to lift economies out of the climate change issue and the normalisation of a corporate approach to carbon management and investment in natural capital offsets (Ervine, 2012; Krahmann, 2012).

While the "common sense" of market environmentalism is evident in the design and rationale of REDD+ (McGregor et al., 2015), the behaviour of PSF actors (investors and companies with shareholders and customers) within it remains relatively unexplored (Bernard et al., 2012). Analysing such behaviour promises to gain insight into their evaluative decisions and capitalist logics regarding how REDD+ projects might satisfy their investment desires (Steurer, 2013). In this way, financialisation (as dynamics of financial motives, markets, actors and institutions influencing the operation of economies) is illustrated where PSF actors collectively influence priorities for financial profits, stock prices and cheap products over "peripheral" REDD+ outcomes, like forest conservation and community development (Knox-Hayes, 2015).

Efforts to revaluate global forests have been complicated by the unstable relationship between PSF actors and the novel REDD+ commodity (Dixon and Challies, 2015). The response and behaviour of PSF has proven heterogeneous, with each foray into REDD+ characterised by different strategies and measures of success (UNEP-FI, 2014) – not least because of REDD+'s inherent contradictions, where nature is "saved" through the same neo-liberal processes facilitating accumulation, volatility and collapse initially jeopardising forests (McAfee, 1999). Further, the uncooperative characteristics of the REDD+ commodity limits its neo-liberalisation, which progresses in diverse ways and at times amounts to greenwashing to disguise private-profit ambitions (Bakker, 2012; Dixon and Challies, 2015).

However, the forest conservation and community development co-benefit aspects of REDD+ commodities offer moral and ethical type values to consumers and markets, and the ways they respond give insight into such value characteristics and market environmentalism prospects (Knox-Hayes, 2015). These co-benefits are an important part of REDD+ projects and are bundled in commodities through REDD+ standards that measure, monitor and report their specific outcomes: from carbon emission reductions, biodiversity conservation, community development and poverty reduction (Lansing, 2012). These REDD+ standards enable (limited) aspects of objectivity in REDD+ projects and fungibility between them, facilitating their exchange (Lansing, 2012; The Munden Project, 2011).

These moral and ethical attributes of REDD+ projects resemble the knowledge and finance type economy that Graham (2006) discusses – specifically that REDD+ demonstrates economies of meaning, extrapolation, evaluation and mediation, where "certain classes of meaning are privileged: . . . there are more and less valuable meanings; . . . access to these meanings is restricted; and . . . meanings can in fact be owned and exchanged, if not entirely consumed" (Graham, 2006, p. 4). Büscher (2014) introduces the idea of epistemic circulation as a way to understand the way meanings and evaluations are negotiated and move between actors on conservation and development networks like REDD+. Channelling Marx, Büscher (2014, p. 87) describes epistemic circulation's recognition of capital as a "value in process" that depends on its persistent circulation in multiple commodity forms, defining epistemic circulation as

> general circulation of interpretations of value through time and space with a particular focus on . . . a specific community of experts sharing a belief in a common set of cause-and-effect relationships as well as common values to which policies governing these relationships will be applied.
>
> (Büscher, 2014, pp. 79–80)

Focusing on the ways that value is a collection of ideas that resonate with certain groups in certain places at certain times creates space to consider both the objective elements of REDD+ projects (as legitimated by REDD+ standards) and the subjective moral and ethical attributes associated with REDD+ conservation and development activities. Together, these objective and subjective aspects of REDD+ projects create knowledges and meanings that are exchanged between actors along REDD+ networks (Dixon and Challies, 2015).

Büscher (2014) recognises how these values can become "fuzzy" over time and when attempts are made to identify their sources within discreet places or agents. These ideas are helpful when conceptualising REDD+ projects as commodities within market environmentalism. The diverse evaluations of REDD+ projects across space and time points to their perceived value beyond the market price of their verified carbon emissions (VERs) (Dixon and Challies, 2015). As Lefebvre (1991, p. 341 emphasis in original) writes, "[t]he commodity is a *thing*: it is *in space* and occupies a location", and thus the REDD+ commodity might also express essential and tangible geographical elements of diversity (Bakker, 2005; Lansing, 2011).

Spatial circuits of value

Understanding REDD+ spatial circuits exposes and enables in-depth analyses of the practices by which actors coalesce and commodify bundles of REDD+ attributes (conservation and enhancement of forest carbon stocks, sustainable management of forests, community development, biodiversity conservation) and "put a value on" them (Coe and Hess, 2011). In this way, they provide insight into how REDD+ projects are commodified into the "DNA of capitalism" as standardised and fungible products to be exchanged at a market price (Bakker, 2005; Watts, 2014).

REDD+ projects within these financialised market contexts experience governing influences from across the desires of investors, companies, customers, shareholders, the public and local actors (Fieldman, 2013). Financialisation's chain of influence is relevant when analysing how the evaluative decisions of distant PSF actors influence Indonesia's forests and REDD+ projects – tracking this influence along project networks and illuminating their motivations and evaluations in and between REDD+ projects. To map these spatial circuits of value, this chapter draws on the GVC framework and SRV theory to explore epistemic circulation within the REDD+ knowledge economy.

Recognising REDD+ projects as discreet intervention packages enables them to be conceptualised as a commodity and open to an analysis of the movement and circulation within globalised value chains (Gereffi and Fernandez-Stark, 2011; Gereffi et al., 1994). The GVC framework facilitates this by exposing the inner workings of "supply chains" and their attributes, components and actors that produce commodities in a specific institutional context (Bernstein and Campling, 2006a, 2006b). The GVC framework facilitates the investigation of global capital's search for value by following the production process of REDD+ projects across space and time. Exploring REDD+ through a GVC lens has been suggested by some authors (Bumpus and Liverman, 2011; Gibbon et al., 2008) but so far not undertaken. While some practitioners have described REDD+ supply chains in a general sense (Bernard et al., 2012), the PhD research that this chapter draws on took the opportunity to explore commodification across three REDD+ project case studies in Indonesia inspired by the GVC framework.

The GVC framework becomes particularly useful in dismantling the processes and politics to illuminate what REDD+ *is* as a commodity, exposing the evaluative practices of PSF and the circulation of value as meanings and knowledges (Büscher, 2014). The traditional treatment of value within GVC analysis has been to understand it as the change in price of the tracked commodity as it passes through a network of actors, as an input and an output (i.e. value added) (Gibbon et al., 2008; Starosta, 2010a; Vagneron and Roquigny, 2011). However, this has recently been disturbed by a reminder of the structural and institutional influence of neo-liberal capitalism on GVCs, complicating neoclassical notions of value as direct subjective price allocation (Gibbon et al., 2008; Selwyn, 2012; Starosta, 2010a, 2010b; Vagneron and Roquigny, 2011). This resonated with the nonlinear meanings, mediations and evaluations witnessed during REDD+ project creation and their relations with the wider institutional context. Subsequently, this analysis looks beyond price-based value-added measures towards SRV theory to help understand the evaluation dynamism in REDD+ projects in Indonesia.

SRV extrapolates an actor's experience with the material things that circulate within exchange networks onto the social relations that construct an actor's knowledges and evaluations of that same material thing (Lee, 2011a). Attention is drawn to the nonlinear feedback relationships between "material circuits of value" (physical things) and the actor knowledges in the places that material things cycle through (Lee, 2011a). SRV lays out the relationship between the motivations and desires of actors (such as those witnessed on REDD+ GVCs in Indonesia) and their actions of evaluation. Evaluation is understood as a social exercise that responds to an actor's place-based knowledges that inform their desires and are shaped by complex and iterative political relationships between their material and social experiences within capitalism. This place-based economic reality resonates closely with Gibson-Graham (2006), who expose the diversity within economies as they are practised in and across places, cementing the importance of relations in and between places when understanding how things are valued.

This link to material circuits of value that drive SRV offers a direct parallel framing for the material focus of inputs and outputs in GVC analysis. In this way, the attribute of value in GVC analysis is open to draw on SRV to help illuminate the gap and add conceptual utility to GVC analysis as a way of understanding commodification and actor behaviour. Lee (2011b) draws attention to this connection, but it is yet to be explored in detail (see Birch [2012] and Foster [2008]). In an effort to add such detail, this chapter contributes an empirically based example of how GVC and SRV can come together. The financialised context that REDD+ projects share as commodities of epistemic circulation position them as particularly exposed to SRV dynamics, given the rapid mobility of knowledge and finance between actors on these REDD+ GVCs in Indonesia.

REDD+ in Indonesia

The emergence of REDD+ in Indonesia has been of major relevance to global forest carbon mitigation efforts because of the country's large forest area, extensive peatlands and high rates of deforestation (Margono et al., 2014). At the same time, palm oil, timber, mining and other extractive land-based industries represent politically and economically powerful, deeply embedded interests. Consequently, Indonesian forests have become ideal candidates for REDD+ finance, and Indonesia has sought to position itself internationally as a REDD+ pioneer. By 2009, the government of Indonesia committed to a national emissions reduction target of 26% below business-as-usual by 2020, or 41% with international support. Indonesia attracted significant international support in 2011, when it signed a Memorandum of Understanding with Norway, which committed US$1 billion in support. In response, Indonesia established a moratorium on new permits to clear primary forests.

By late 2011, the national REDD+ Task Force was established to implement the moratorium, oversee Indonesia's REDD+ programme and launch the National REDD+ Strategy in 2012. Meanwhile, Indonesia has continued to progress its national REDD+ readiness and finance efforts, during which between 44 and 77 REDD+ projects and provincial REDD+ pilots were developed (Minang et al., 2014). The research this chapter is based on gathered evidence from interviews and extended field research conducted in Indonesia between April 2013 and March 2014. Key informants include actors from across the REDD+ sector, encompassing international and Indonesian government institutions, public and private sector financiers, project developers, businesses, extractive industry, consultants and contractors, international and local NGOs and forest community members.

PSF investment in REDD+ projects for financial returns has been receding and replaced by a smaller but potentially expanding body of finance pursuing less financial profit and with more complicated desires (Dixon and Challies, 2015; GCP et al., 2014). These organisations are influenced by financialisation, but they tend to seek investment opportunities with "ethical" or "responsible" attributes to incorporate into their organisational reporting to shape their organisational identity – commonly referred to as corporate social responsibility (CSR). For these actors, the price of the carbon offset is not the primary factor determining their interest but rather the various meanings attached to carbon take on more significance. The declining relative importance of carbon offsets, their ability to deliver financial profits and the increasing pursuit of CSR "stories" that carry meaning has undermined the fungibility of REDD+ projects, as they attempt to accommodate and respond to the particular tastes of PSF actors (The Munden Project, 2011).

The characteristic REDD+ GVC took the form of PSF actors (like an investor or company) financing a REDD+ project developer in Indonesia (typically an NGO) to be responsible for working with forest communities, authorities and REDD+ standards to develop and secure inputs into the REDD+ project and CSR story. These project developers inhabited an influential position on these GVCs and were able to modify the inputs and story of the REDD+ project as perceived by PSF actors, to best suit and attract them. These arrangements to develop REDD+ projects constituted the input–output chains of GVCs and the material circuits of value of SRV theory.

Diverse motivations, knowledges and stories of REDD+ GVCs

The analysis showed that actors in REDD+ project GVCs each contributed and responded to meanings and knowledges in unique and diverse ways. The motivations and desires of PSF actors were indicated by the reasons why they might fund a project that pursued certain meanings and knowledges in its story (Dixon and Challies, 2015). Given the financialised capitalist context that these PSF actors rest in, such actions can be considered as reflective of the respective or prospective desires of their neighbouring GVC members – concerning forest conservation, community development and financial profit.

One informant described how REDD+

> Brought all these different people together who all thought that there was something in it for each of them.
>
> (Contractor 2)

The heterogeneity of motivation and desire on GVCs in each REDD+ project complicated the fungibility, financialisation, and thus marketability of forest carbon activities:

> There are lots of different reasons why people are doing these things and it's constantly weighing things up; reputational issues vs moral issues vs having to make a commercial return.
>
> (INGO 2)

These different motivations of PSF in REDD+ projects demonstrate the mixture of knowledges and evaluations of different actors from different places along the project GVC.

The plurality in motivations and desires between REDD+ actors on GVCs was illustrated among the complex diversity of REDD+ projects: "It's going to be done a zillion different ways around the world, because it has to be" (INGO 2). PSF actors sought REDD+ projects that were legitimised by REDD+ standards with stories consistent with the meanings they wanted associated with their organisation, thus evaluating them as desirable. However, it was recognised that these knowledges and meanings that became significant sources of value in projects could become difficult to locate and define:

> You can always find additionality with REDD+ if you fish around for it. The additionality can get a bit funky after a while.
>
> (Consultant 1)

The informant is referring to the way that the additional benefits the project delivers can be justified through relatively abstract sources that can justify the creation and re-creation of stories, knowledges and meanings to suit the actors involved. By doing so, the stories created on REDD+ GVCs reflect the characteristics of knowledge commodities – composed of privileged meanings that change through time, becoming detached from the standardised and "objective" place-based measures that initially justified them.

This reflexive relationship between multiple PSF desires and the diverse stories of REDD+ projects was evident in the way projects had accommodated the interests of PSF actors. One informant described the influence of finance on REDD+ project developers:

> Essentially, they've written a proposal depending on that finance. They've said they would deliver something that's based on the values of that funding institution.
>
> (Consultant 1)

The specific characteristics of PSF desires came to influence REDD+ projects to produce and privilege certain meanings and knowledges. Here the knowledges of PSF actors within financial rationales were demonstrably influential among the social relations of value along REDD+ GVCs.

The GVC framework exposed the epistemic circulation of the REDD+ knowledge commodity, showing how the meanings and knowledges created in projects constituted the stories that PSF actors (and their sources of financial pressure) desired association with. Subsequently, meanings within REDD+ knowledge commodities influenced the types of PSF actors involved, which in turn influenced the character of the REDD+ projects and their knowledge commodities, reflecting the reflexive dynamic of SRV theory.

Evaluation among REDD+ project GVCs in Indonesia

The place specific anchoring of SRV resonates strongly with the observation of different PSF actors with different knowledges influencing, and influenced by, REDD+ projects. The reflexive dynamism of social and material experiences in SRV is demonstrated by PSF actors' perpetually revaluating their position in REDD+ projects. PSF actors are increasingly seeking ethical and sustainable standards and stories in their products and portfolios and their financial pressure influences the other actors on their networks to seek out REDD+ knowledge commodities. This has exposed how the diversity of PSF actors place-based knowledges can shape the REDD+ programme and offers aspiration and reflection of "a vision of global transformation through the accretion and interaction of small changes in place" (Gibson-Graham, 2006, p. 196). The rapidly emergent and experimental nature of REDD+ projects presented opportune ground to explore this relationship between GVC analysis and SRV.

Doing so embraces the value multiplicity evident in and between PSF actors and provides helpful insight into approaching value attributes in GVC and commodity analyses. SRV brought together actions of financial accumulation and ecological conservation in REDD+ by recognising subjective knowledges and their reflexive influences on evaluation. In this way, SRV can complement and help understand the character of GVCs, building on the input–output approach to networks with nuanced and socially dynamic circuits of value.

Furthermore, this hints that the challenge for REDD+ may be less about putting a value on nature (like an international carbon price) and more about influencing social constructions of value, evaluative practices and knowledges in diverse economies to engage questions of ecological limits and their reflections on what is desirable. And they may have to do this in much the same way that has been witnessed among the PSF actors of Indonesia's REDD+ knowledge commodities, who's shareholders and consumers have been asking, are starting to ask, or are expected to ask such questions.

References

Anderson, T.L. and Leal, D.R. (2001) *Free Market Environmentalism*. New York: Palgrave Macmillan.

Bakker, K. (2005) 'Neoliberalizing Nature? Market Environmentalism in Water Supply in England and Wales', *Annals of the Association of American Geographers*, 95(3), pp. 542–565.

Bakker, K. (2012) 'The "Matter of Nature" in Economic Geography', in Barnes, T.J, Peck, J. and Sheppard, E. (eds) *The Wiley-Blackwell Companion to Economic Geography*, West Sussex: Wiley-Blackwell, pp. 104–117.

Bernard, F., McFatridge, S. and Minang, P. (2012) *The Private Sector in the REDD+ Supply Chain: Trends, Challenges and Opportunities*. Winnipeg, Canada: IISD.

Bernstein, S. and Campling, L. (2006a) 'Commodity Studies and Commodity Fetishism I: Trading Down', *Journal of Agrarian Change*, 6(2), pp. 239–264.

Bernstein, S., and Campling, L. (2006b) 'Commodity Studies and Commodity Fetishism II: "Profits With Principles"?' *Journal of Agrarian Change*, 6(3), pp. 414–447.

Birch, K. (2012) 'Knowledge, Place, and Power: Geographies of Value in the Bioeconomy', *New Genetics and Society*, 31(2), pp. 183–201.

Bumpus, A.G. and Liverman, D. (2011) 'Carbon colonialism? Offsets, greenhouse gas reductions and sustainable development', in Peet, R., Robbins, P. and Watts, M. (eds) *Global Political Ecology*. Oxon, New York: Routledge, pp. 203–224.

Büscher, B. (2014) 'Selling Success: Constructing Value in Conservation and Development', *World Development*, 57, pp. 79–90.

Coe, N. M., and Hess, M. (2011) 'Local and Regional Development: A Global Production Network Approach', in Pike, A., Rodriguez-Pose, A. and Tomaney, J. (eds) *Handbook of local and regional development*, Oxford and New York: Routledge, pp. 128–138.

Dixon, R. and Challies, E. (2015) 'Making REDD+ pay: Shifting Rationales and Tactics of Private Finance and the Governance of Avoided Deforestation in Indonesia', *Asia Pacific Viewpoint*, 56(1), pp. 6–20.

Dunlap, A.A. (2015) 'The Expanding Techniques of Progress: Agricultural Biotechnology and UN-REDD+', *Review of Social Economy*, 73(1), pp. 89–112.

Ervine, K. (2012) 'The Politics and Practice of Carbon Offsetting: Silencing Dissent', *New Political Science*, 34(1), pp. 1–20.

Ervine, K. (2013) 'Diminishing Returns: Carbon Market Crisis and the Future of Market-Dependent Climate Change Finance', *New Political Economy*, 19(5), pp. 723–747.

Fankhauser, S. (2010) 'The Costs of Adaptation', *Wiley Interdisciplinary Reviews: Climate Change*, 1(1), pp. 23–30.

Fieldman, G. (2013) 'Financialisation and Ecological Modernisation', *Environmental Politics*, 23(2), pp. 224–242.

Foster, R.J. (2008) 'Commodities, Brands, Love and Kula: Comparative Notes on Value Creation in Honor of Nancy Munn', *Anthropological Theory*, 8(1), pp. 9–25.

GCP, IPAM, FFI, UNEP FI and UNORCID (2014) *Stimulating Interim Demand for REDD+ Emission Reductions: The Need for a Strategic Intervention from 2015 to 2020*, Global Canopy Programme, Oxford, UK; the Amazon Environmental Research Institute, Brasília, Brazil; Fauna & Flora International, Cambridge, UK; UNEP Finance Initiative, Geneva, Switzerland; and United Nations Office for REDD+ Coordination in Indonesia, Indonesia.

Gereffi, G. and Fernandez-Stark, K. (2011) *Global Value Chain Analysis: A Primer. Center on Globalization, Governance & Competitiveness,* Durham, NC: Center on Globalization, Governance & Competitiveness (CGGC), Duke University.

Gereffi, G., Korzeniewicz, M. and Korzeniewicz, R. (1994) 'Introduction: Global Commodity chains', in Gereffi, G. and Korzeniewicz, M. (eds) *Commodity Chains and Global Capitalism*. Westport, CT: Praeger, pp. 1–14.

Gibbon, P., Bair, J. and Ponte, S. (2008) 'Governing Global Value Chains: An Introduction', *Economy and Society*, 37(3), pp. 315–338.

Gibson-Graham, J.K. (2006) *A Postcapitalist Politics*. London; Minneapolis, MN: University of Minnesota Press.

Graham, P. (2006) *Hypercapitalism: Language, New Media, and Social Perceptions of Value*. New York: Peter Lang.

Hiraldo, R. and Tanner, T. (2011) *The Global Political Economy of REDD+: Engaging Social Dimensions in the Emerging Green Economy*. Geneva, Switzerland: United Nations Research Institute for Social Development.

Knox-Hayes, J. (2015) 'Towards a Moral Socio-Environmental Economy: A Reconsideration of Values', *Geoforum*, 65, pp. 297–300. doi:10.1016/j.geoforum.2015.07.028

Krahmann, E. (2012) 'Green Consumer Markets in the Fight Against Climate Change', *European Security*, 22(2), pp. 1–18.

Lansing, D.M. (2011) 'Realizing Carbon's Value: Discourse and Calculation in the Production of Carbon Forestry Offsets in Costa Rica', *Antipode*, 43(3), pp. 731–753.

Lansing, D.M. (2012) 'Performing Carbon's Materiality: The Production of Carbon Offsets and the Framing of Exchange', *Environment and Planning A*, 44(1), pp. 204–220.

Lee, R. (2011a) 'Spaces of hegemony? Circuits of value, finance capital and places of financial knowledge', in Agnew, J. and Livingstone, D.N (eds) *The SAGE Handbook of Social Geographies*. London: SAGE Publications Ltd, pp. 185–202.

Lee, R. (2011b) 'Within and outwith/material and political? Local Economic Development and the Spatialities of Economic Geographies', in Pike, A., Rodriguez-Pose, A. and Tomaney, J. (eds) *Handbook of Local and Regional Development*. Oxford, New York: Routledge, pp. 193–211.

Lefebvre, H. (1991) *The Production of Space. Production*. Oxford: Wiley-Blackwell.

Lövbrand, E. and Stripple, J. (2012) 'Disrupting the Public – Private Distinction: Excavating the Government of Carbon Markets Post-Copenhagen', *Environment and Planning-Part C*, 30(4), pp. 658–674.

Margono, B. A., Potapov, P. V, Turubanova, S., Stolle, F., and Hansen, M. C. (2014) 'Primary Forest Cover Loss in Indonesia Over 2000–2012', *Nature Climate Change*, 4, pp. 730–735.

McAfee, K. (1999) 'Selling Nature to Save It? Biodiversity and Green Developmentalism', *Environment and Planning D: Society and Space*, 17(2), pp. 133–154.

McGregor, A., Challies, E., Howson, P., Astuti, R., Dixon, R., Haalboom, B., . . . Afiff, S. (2015) 'Beyond Carbon, More Than Forest? REDD+ Governmentality in Indonesia', *Environment and Planning A*, 47(1), pp. 138–155.

Minang, P. A., Van Noordwijk, M., Duguma, L. a, Alemagi, D., Do, T. H., Bernard, F., . . . Leimona, B. (2014) 'REDD+ Readiness Progress Across Countries: Time for Reconsideration', *Climate Policy*, 14(6), pp. 685–708.

The Munden Project. (2011) *REDD and Forest Carbon: Market-Based Critique and Recommendations*. Available at: www.rightsandresources.org/documents/files/doc_2215.pdf?.

Nakhooda, S., Fransen, T., Kuramochi, T., Caravani, A., Prizzon, A. et al. (2013) *Mobilising International Climate Finance: Lessons from the Fast-Start Finance Period*. London: ODI.

OECD. (2011) *Towards Green Growth*. Paris: OECD Green Growth Studies, OECD Publishing.

Paterson, M. (2012) 'Who and What Are Carbon Markets for? Politics and the Development of Climate Policy', *Climate Policy*, 12(1), pp. 82–97.

Paterson, M., Hoffmann, M., Betsill, M. and Bernstein, S. (2013) 'The Micro Foundations of Policy Diffusion Toward Complex Global Governance: An Analysis of the Transnational Carbon Emission Trading Network', *Comparative Political Studies*, 47(3), pp. 420–449.

Peters-Stanley, M., and Yin, D. (2013) *State of the Voluntary Carbon Markets 2013: Maneuvering the Mosaic*. Washington, DC: Forest Trends' Ecosystem Marketplace and Bloomberg New Energy Finance.

Rogers, P.P., Jalal, K.F. and Boyd, J.A. (2008). *An Introduction to Sustainable Development* (Vol. 24). London: Earthscan.

Selwyn, B. (2012) 'Beyond Firm-Centrism: Re-integrating Labour and Capitalism into Global Commodity Chain Analysis', *Journal of Economic Geography*, 12(1), pp. 205–226.

Starosta, G. (2010a) 'Global Commodity Chains and the Marxian Law of Value', *Antipode*, 42(2), pp. 433–465.

Starosta, G. (2010b) 'The Outsourcing of Manufacturing and the Rise of Giant Global Contractors: A Marxian Approach to Some Recent Transformations of Global Value Chains', *New Political Economy*, 15(4), pp. 543–563.

Steurer, R. (2013) 'Disentangling Governance: A Synoptic View of Regulation by Government, Business and Civil Society', *Policy Sciences*, 46(4), pp. 387–410.

Sukhdev, P. (2011) Put a value on nature! Retrieved June 3, 2015, from www.ted.com/talks/pavan_sukhdev_what_s_the_price_of_nature

UNEP-FI. (2014) *Demistifying Private Climate Finance*. Geneva, Switzerland: UNEP – Finance Initiative.

UNFCCC. (2007) *Investment and Financial Flows to Address Climate Change*. Bonn, Germany: United Nations Framework Convention on Climate Change.

UN-REDD. (2015) *The UN-REDD Programme Strategy 2011–2015*. Geneva, Switzerland: UN-REDD Programme. Available at: www.unredd.net/index.php?option=com_docman&task=doc_download&gid=4598&Itemid=53.

Vagneron, I. and Roquigny, S. (2011) 'Value Distribution in Conventional, Organic and Fair Trade Banana Chains in the Dominican Republic', *Canadian Journal of Development Studies/Revue Canadienne D'études Du Développement*, 32(3), pp. 324–338.

Watts, M. (2014) 'Commodities', in Cloke, P., Crang, P. and Goodwin, M. (eds) *Introducing Human Geographies* (3rd ed.). New York: Routledge, pp. 391–412.

Index

Printed in the United States
by Baker & Taylor Publisher Services

Printed in the United States
by Baker & Taylor Publisher Services